Optimization in the Agri-Food Supply Chain

Optimization in the Agri-Food Supply Chain

Recent Studies

Edited by

Mayssa Koubaa
Mohamed Haykal Ammar
Diala Dhouib
Sirine Mnejja

ISTE

WILEY

First published 2024 in Great Britain and the United States by ISTE Ltd and John Wiley & Sons, Inc.

ISTE Ltd
27-37 St George's Road
London SW19 4EU
UK

www.iste.co.uk

John Wiley & Sons, Inc.
111 River Street
Hoboken, NJ 07030
USA

www.wiley.com

Library of Congress Control Number: 2024938140

British Library Cataloguing-in-Publication Data
A CIP record for this book is available from the British Library
ISBN 978-1-78630-809-2

Contents

Preface

In recent years, the optimization of supply chains in the agri-food and animal industries has been a topic of significant research and study. With the increasing complexity and challenges faced by these industries, finding efficient and sustainable solutions has become crucial. In this context, technology integration explores how emerging technologies such as the Internet of Things (IoT), blockchain and artificial intelligence (AI) can be integrated into the supply chain management of industries. Furthermore, the sustainability focus emphasizes the importance of viable practices in supply chain optimization. It explores how incorporating sustainability metrics into decision-making processes can lead to reducing environmental impact and increasing social responsibility. This could include considering factors such as carbon emissions, water usage and ethical sourcing practices, when designing and managing supply chains. In addition to these concepts, this book addresses the need for resilient supply chains in the face of unforeseen events and disruptions, such as natural disasters or disease outbreaks. It explores strategies for risk management, including diversifying sourcing locations, implementing contingency plans and using technology to quickly adapt to changing circumstances.

This book aims to provide an overview of recent studies conducted in the field of agri-food supply chain optimization and animal industry. It also explores the impact of climate change on the food–energy–water nexus, with a focus on Poland as a case study. Additionally, this book discusses the management and analysis of sustainable initiatives in the agri-food supply chain in India, inventory management activities in the Indian sugar supply chain, the dynamics of the rice supply chain system in Punjab, Pakistan and the culture and use of the elderberry bush for the production of elderberry jam. Furthermore, it will delve into the application of the strength Pareto evolutionary algorithm to solve multi-objective transportation problems, the comparison of classical and machine learning forecasting techniques in rice production in India, the prioritization of circular production of olive oil in Tunisia and the role of open innovation as a solution in the food supply chain.

The agri-food supply chain is a complex network of activities involved in the production, processing and distribution of agricultural products. The optimization of this supply chain is essential to ensure efficiency, reduce costs and enhance sustainability. In this context, the first chapter emphasizes the selection of tomato varieties in agri-food supply chain using Delphi–AHP (Analytic Hierarchy Process) methodology and MOORA–MULTIMOORA approaches for an Algerian case study. The tomato variety selection problem (TVSP) is an important decision in agri-food supply chain management. It involves choosing the best tomato variety from a set of alternatives using a set of evaluation criteria. The Delphi method is a formal consensus approach to identifying a problem, selecting panelists, developing a questionnaire and determining whether consensus is reached. The AHP method is considered as a popular multi-criteria decision-making method and is known for its ease of application and efficiency. The problem of tomato variety selection in the agri-food supply chain has not been addressed in previous studies, so the objective of this research is to integrate Delphi, AHP, MULTIMOORA and MOORA approaches to solve this problem. The application of the proposed methodology to the tomato processing company in Algeria, including these three methodologies, the findings and the analysis of the implications of this study are discussed in this chapter.

The second chapter presents an analysis of the current trends in climate change and its effects specifically in Central Europe, especially in Poland due to its vulnerability to water shortages. Climate change has become a significant global concern, impacting various sectors, including agriculture and food production. The main focus of the analysis is on the food–energy–water nexus, which refers to the interconnectedness and interdependence of these three essential resources. In this chapter, the authors discuss the main challenges and opportunities that Poland faces in relation to the food–energy–water nexus. Moreover, they identify knowledge gaps and suggest directions for future research that could lead to sustainable, cost-effective and feasible solutions. These solutions would be applicable not only to Poland but also to other countries in the region that face similar challenges.

Chapter 3 discusses a study that aims to identify sustainable initiatives in the agricultural food product business. It includes a literature review and inputs from domain specialists to identify these initiatives that focus on optimizing environmental and socio-economic factors. In this chapter, the authors prove that these initiatives have a significant impact on the agricultural food product business. In order to analyze these initiatives, researchers use the fuzzy analytic hierarchy process (FAHP) approach. This approach is a decision-making method that allows consideration of multiple criteria and their relative importance. By applying the FAHP approach, it is possible to provide findings that have managerial and policy implications for key stakeholders.

Chapter 4 describes the objectives and methodology of a study on inventory management activities in the sugar industry. Inventory management plays a crucial role in ensuring the smooth functioning of supply chains, including the Indian sugar supply chain. This chapter aims to identify issues related to inventory management and explores the impact of IT applications in facilitating inventory management. The first part of the chapter identifies relevant issues associated with managing inventory activities. In addition, the second part involves the application of factor analysis to determine the factors that govern inventory management activities in the sugar industry. Furthermore, researchers use the regression method to establish a relationship between these issues and the facilitation provided by IT applications. As a final step, the hypotheses are tested using ANOVA to determine whether the relationships observed in the regression analysis are statistically significant.

Chapter 5 discusses the importance of rice as the second most important cereal crop in Pakistan, with a significant portion consumed domestically and the rest exported to the Gulf countries. Rice is a staple crop in Punjab, Pakistan and its supply chain system is of significant importance to the region's economy. This chapter aims to understand the dynamics and market margins of the rice value chain in domestic markets, and to identify barriers and opportunities for sustainable development. Data was collected through questionnaires and focus group interviews with key actors in the rice value chain. The analysis revealed problems such as a lack of extension services, infrastructure and marketing information systems, which contribute to vulnerability in the chain. Commission agents and wholesalers were found to have the largest share of the marketing margin. The main highlighted parts of this chapter are the need for advanced production technologies, skilled labor, input subsidies and the development of micro-institutions to support smallholder farmers in a sustainable manner.

Chapter 6 focuses on the four-dimensional, multi-objective transportation problem in the General Company for Food Stuff Trading. Transportation plays a critical role in the agri-food supply chain, ensuring the timely and efficient delivery of goods. This chapter aims to manage the transportation process by considering transportation costs, time and distance for different demand and supply centers, distribution lines and means of transportation. The researchers use the strength Pareto evolutionary algorithm and MATLAB program to solve the transportation model and obtain the lowest transportation cost, time and distance. The objectives of this chapter include using realistic applied cases, integrating AI algorithms, speeding up program execution using the decoding procedure and analyzing the homogeneity of objectives in the four-dimensional model using the Pareto group.

Chapter 7 investigates various forecasting methods, including classical methods and machine learning techniques, to predict agricultural rice production in India. Accurate forecasting of rice production is crucial for effective planning and decision-making in the agricultural sector. The time series data from 1949 to 2021 is used for the analysis. Performance metrics such as MAPE (Mean Percentage Error), RMSE (Root Mean Square Error) and MAE (Mean Absolute Error) are used to compare the forecasting methods. The findings prove that the random forest model outperforms other models and predicts a value of 108892.7 thousand tons for rice production in 2022.

Chapter 8 aims to help decision-makers choose sustainable agricultural practices, transformation methods and waste management for the life cycle of the olive oil chain. Circular production, which minimizes waste and maximizes resource efficiency, has become a key concept in sustainable development. By applying circular economy principles and using life cycle analysis methods, different scenarios can be identified and compared to determine the most sustainable option. The study focuses on evaluating the sustainability of existing olive systems in Tunisia, based on eight alternatives proposed in the scientific literature.

Chapter 9 explores the role of open innovation in addressing the challenges faced by the food supply chain. Open innovation, which involves the collaboration and exchange of ideas between different stakeholders, has emerged as a solution in the food supply chain. In a rapidly changing economic environment, food supply chains are facing challenges due to globalization, lower trade barriers and various crises. To promote sustainable growth, businesses are adopting specific management practices that encourage innovation. This has led companies to collaborate with external partners such as customers, suppliers, universities and stakeholders. The motivation behind this is to promote economic development in the agri-food sector and ensure its further growth through open innovation.

Chapter 10 deals with agri supply chain optimization and proposes a Mixed Integer Non-Linear Programming that aims to optimize cost with emissions, considering inventory and safety stocks, and leveraging blockchain technology in the case study of the Islamic Republic of Iran.

In conclusion, this book aims to provide a comprehensive overview of recent studies conducted in the field of agri-food supply chain optimization and animal industry. It explores various topics, including the impact of climate change on the food–energy–water nexus, sustainable initiatives in the agri-food supply chain, inventory management activities, the dynamics of specific supply chain systems, the culture and use of underutilized crops, optimization techniques, forecasting methods, circular production and open innovation. By examining these areas, this book contributes to the body of knowledge on supply chain optimization and offers insights and recommendations for enhancing efficiency and sustainability in the agri-food and animal industries.

Acknowledgments

This book is supported by the Tunisian Ministry of Higher Education and Scientific Research and in collaboration with the "Partnership for Research and Innovation in the Mediterranean Area" (PRIMA) initiative.

Mayssa Koubaa
Mohamed Haykal Ammar
Diala Dhouib
Sirine Mnejja
June 2024

1

Selection of Tomato Varieties on Agri-Food Supply Chain: Delphi–AHP Methodology and MOORA–MULTIMOORA Approaches for an Algerian Case Study

1.1. Introduction

The tomato variety selection problem (TVSP) presents one of the most relevant strategic decisions, which will participate efficiently in the strategic planning of the management of the agri-food supply chain. It significantly influences the quality of the final product and the whole supply chain's performance. The TVSP is a decision activity, which ensures the best choice of the potential tomato variety among a set of alternatives. A set of criteria must be used to evaluate the different varieties.

Multi-criteria decision-making (MCDM) methods present the most relevant part of the operational research field, which can be used to deal with complex and multiple objective problems. Moreover, several targeted objectives in the same problem can put the decision-maker in an ambiguous situation, and hence the selection between available alternatives will be more complex. MCDM methods could provide an explicit tradeoff between several criteria in order to select the most appropriate options among the feasible solutions (Aramyan et al. 2007).

1.2. Literature review

The Delphi is a formal consensus approach and a systematic means of measuring and developing consensus among partners (Humphrey-Murto and de Wit 2019). It is applied to several domains such as medicine, manufacturing, economy, etc. The

Chapter written by Mohamed Amine CHERIER and Amina BENDJELLOUL.

Delphi method includes identifying the problem to be solved, selecting panelists, developing a questionnaire that must be sent to experts to collect individual and group opinions and, finally, determining if a consensus is reached (Emovon et al. 2018). Bui et al. (2020) applied the fuzzy Delphi method to acquire valid and reliable attributes through qualitative information; the results show 44 essential barriers to sustainable solid waste management. Abreu and Mesias (2020) presented a set of demographic, economic, environmental and social welfare indicators that have to be taken into account in order to evaluate rural development and related policies using the Delphi method.

The analytic hierarchy process (AHP) method is considered one of the famous MCDM methods, as it is easy to apply and efficient in terms of results. Giamalaki and Tsoutsos (2019) presented and tested a methodology for clarifying and prioritizing the most suitable locations for siting solar power installations using Geographical Information Systems and the AHP methodology. Mukeshimana et al. (2021) presented a study that applied a combination of both qualitative and quantitative analyses to identify and analyze the major constraints that hindered the dissemination of domestic biogas in Rwanda. The results showed that the financial category is the most influential barrier, with high capital cost and lack of a financial mechanism highly ranked among financial, technical, socio-cultural and institutional barriers. Dožić et al. (2018) presented, in their paper, a methodology that will help choose aircraft type(s) that meets the market conditions and an airline's requirements for the best known route network and known forecasted air travel demand by routes using the fuzzy analytic hierarchy process (FAHP) method.

Patnaik et al. (2020) used the AHP–MOORA method, which is a hybrid MCDM approach used to select materials for good wear resistance and a structural application's point of view. The selection processes are carried out by considering the physical, mechanical and wear properties of the composite materials such as density, hardness, tensile strength, etc. Various factors affecting the service quality of customers have been identified and ranked with the help of two approaches of multi-objective optimization, based on the ratio analysis (MOORA) approach (Gupta et al. 2022).

According to the literature presented, the problem of tomato variety selection has not been solved, although it has importance in the production and world tomato security. It is well known that despite a large number of studies integrating different MCDM methods for the agri-food supply chain, there is no study that has integrated Delphi, AHP, MULTIMOORA and MOORA approaches as methodological solutions to solve the problem of tomato variety selection.

To fulfill this research gap, the objective of this research is to integrate Delphi–AHP–MOORA–MULTIMOORA for the TVSP in the agri-food supply chain. The proposed integrated approach has been applied to select tomato varieties at a tomato processing company in Algeria.

1.3. Methodology

1.3.1. *Delphi method*

The Delphi method is an investigative approach, which is used to collect, modify and analyze an expert's opinions, aiming to achieve a consensus in order to solve complex decision problems (Humphrey-Murto and de Wit 2019). The Delphi methodology has been considered as a powerful method to bring out the opinions of experts on the security food domain (Wentholt et al. 2010). Moreover, it is an effective method for selection criteria based on the opinions of experts (Kamali et al. 2015). The application of this approach involves two main steps to follow. Firstly, clearly identify the problem that would be addressed. Subsequently, select a panel of experts, who will participate with their knowledge and experiences. Develop a questionnaire survey that will be sent to anonymous experts in order to obtain their opinions (the first round of the Delphi questionnaire). The second step is processing and analyzing the experts' responses. Based on the result of the first step, the researcher or investigator sends a new questionnaire survey (which is mostly a modified one) to the team of experts for another revision (the second round of the Delphi questionnaire). The second round must be repeated until the consensus is reached, generally achieved in the second or third round (Emovon et al. 2018). It is worth mentioning that the anonymity is indispensable, which is characteristic of the Delphi method to achieve a good result.

The proposed methodology solution includes several advantages. Firstly, every selected expert has the authority to be included in any part of the Delphi procedure. Secondly, any panelist can freely express their opinion without the need for argumentation (Kim et al. 2013). Thirdly, regarding other data-gathering methods, the Delphi approach leads to the consensus of experts. This proposal has been considered the best methodology for brainstorming policy.

1.3.2. *Analytic hierarchy process (AHP) method*

The AHP is one of the MCDM methods. It is a decision-making tool used for complicated and unstructured problems. It is a powerful approach that facilitates decision-making with the organization of judgments, feelings and understandings into an organized hierarchical structure in order to make a coherent decision. It is characterized by the decomposition of a complex problem into a multilevel hierarchy structure: final objectives, criteria, sub-criteria and alternatives (Saaty 2008). The AHP method has proved its flexibility and its capability to solve complex problems with multi-criteria on different fields. It is considered as an efficacy tool that aids decision-makers to structure different components of a problem into a family tree in order to facilitate making decisions (Ni-Di and Ming-Xian 2010). The AHP can combine both qualitative and quantitative factors to rank feasible solutions (Saaty 2008; Wang 2010). Although there are numerous methods proposed in the literature for solving multi-criteria decision problems, the AHP remains the better tool for dealing with this type of problem due to its applicability in multiple domains and ease of use (Harputlugil et al. 2011).

1.3.2.1. *AHP process*

As mentioned above, the AHP is a methodological approach that can give the decision-maker good support in the conflicting problem. It is an easy and important method to assess and rank alternatives to complex and unstructured problems by assigning weights.

The AHP method can be presented in three main phases. These phases are (1) identifying the structure of the hierarchy; (2) pairwise comparison; and (3) making the best decision among the alternatives (Oguztimur 2011).

For the first phase, the identification of the hierarchy structure is required by selecting the criteria, sub-criteria and alternatives.

The second phase is the formation of the pairwise assessment matrix in order to evaluate the importance of each factor. The experts and decision-makers use the numerical values indicated in the Saaty scale (Table 1.1) to evaluate every pairwise comparison.

Values	Explanation of values
1	Equal importance of two factors
3	Weak preference (factor i over factor j)
5	Strong preference (i over j)
7	Very strong important (i over j)
9	Extreme preference (i over j)
2, 4, 6, 8	Intermediate values between two judgments (i over j)

Table 1.1. *Pairwise comparison scale (Saaty 2008)*

The result of the pairwise comparison should be evaluated by experts to obtain the matrix M:

$$M(m_{ij}) = \begin{bmatrix} 1 & m_{12} & m_{13} & \cdots & m_{1n} \\ m_{21} & 1 & m_{23} & \cdots & m_{2n} \\ \vdots & \vdots & \vdots & \cdots & \vdots \\ \vdots & \vdots & \vdots & \ddots & \vdots \\ m_{n1} & m_{n2} & m_{n3} & \cdots & 1 \end{bmatrix} \quad m_{ii} = 1,\ m_{ji} = 1/m_{ij},\ m_{ij} \neq 0 \qquad [1.1]$$

M is a comparison matrix that will take the number 1 in its diagonal and (m_{ij}) number for the pairwise comparison between the elements i and j, (i, j = 1, 2, 3, . . . n).

When the comparison matrix is formed, the calculation of the weight vector W can be determined by solving equation [1.2], where λ_{max} is the maximum average eigen value (Joshi et al. 2011). The principal right eigenvector method is used to determine the weight vector of the pairwise comparison matrix (Saaty 2008).

$$MW = \lambda_{max} \otimes W \qquad [1.2]$$

The deciders should then verify the consistency of the result by calculating the CR factor. CR is a detected mean that enables the deciders to ensure the reliability of the pairwise comparison. We can effectively calculate it by the mathematical equation [1.3]:

$$CR = CI/RI \qquad [1.3]$$

- RI: the random consistency index (Table 1.2).

- CI: Consistency Index, determined by equation [1.4]:

$$CI = (\lambda_{max} - n)/n - 1 \qquad [1.4]$$

- n: the number of rows in the judgment matrix.

n	1	2	3	4	5	6	7	8	9	10	11	12	13	14	15
RI	0.00	0.00	0.52	0.89	1.12	1.25	1.35	1.40	1.45	1.49	1.52	1.54	1.56	1.58	1.59

Table 1.2. *Random consistency index according to the number of n (Saaty 2008)*

The judgment matrix is consistent if the CR is inferior to 0.10. Therefore, the result of the matrix values is considered coherent. Otherwise, the experts are summoned to consider another revision to reduce the inconsistencies (Saaty 2008).

The third phase, which is the final step, aims to select the best alternatives after analyzing and an assessment of the result.

1.3.3. *The MOORA method*

MOORA represents a strategy for multi-criteria optimization. It tries to meet a number of requirements, including unit selection for each objective, normalization, optimization and value (Brauers and Zavadskas 2006). It is additionally used to obtain the most effective solution from a collection of discrete alternatives with opposing objectives. It has been used to solve a number of economic, management and construction issues (Kuram and Ozcelik 2013; Maiyar et al. 2013; Kivak 2014; Shihab and Mubarak 2017).

The development of a decision matrix is the first step. The decision matrix depicts the relative output of various alternatives based on various criteria:

$$M(m_{ij}) = \begin{bmatrix} x_{11} & x_{12} & x_{13} & \cdots & x_{1n} \\ x_{21} & 1 & x_{23} & \cdots & x_{2n} \\ \vdots & \vdots & \vdots & \cdots & \vdots \\ \vdots & \vdots & \vdots & \ddots & \vdots \\ x_{m1} & x_{m2} & x_{m3} & \cdots & x_{mn} \end{bmatrix} \qquad [1.5]$$

x_{ij} presents the performance value of ith alternative on ij^{th}criterion, and m and n are the numbers of alternatives and criteria, respectively.

The move that makes the output indices dimensionless and comparable is the normalization of the decision matrix, also known as a ratio system. Each alternative's performance value on a criterion is calculated by comparing it to the other alternative's performance on that criterion:

$$x_{ij}^{*} = \frac{x_{ij}}{\sqrt{\sum_{i=1}^{m} x_{ij}^{2}}} \qquad [1.6]$$

x_{ij}^{*} is a dimensionless number between [0,1] and the normalized performance of ith alternative on j^{th} criterion.

The normalized performances are added for beneficial criteria and subtracted for non-beneficial criteria [1.7]. The result is the overall performance score of each alternative (y_i^*).

$$y_{ij}^{*} = \sum_{i=1}^{n} x_{ij}^{*} - \sum_{i=1}^{g+1} x_{ij}^{*\,2} \qquad [1.7]$$

In this formula, g and (n-g) are the number of criteria to be maximized and minimized, respectively. The y_j^* value can be positive or negative depending on the total of its maxima (beneficial attributes) and minima (non-beneficial attributes) in the decision matrix. An ordinal ranking of y_j^* shows the final preference. As a result, the best alternative has the highest y_j^* value, while the worst alternative has the lowest y_j^* value.

1.3.3.1. *The full multiplicative form MULTIMOORA*

The full multiplicative form of multiple criteria consists of both maximization and minimization of a purely multiplicative utility function. The main characteristics of this form are being nonlinear, non-additive and not using attribute weights. Overall, the utility of the ith alternative (u_j) is computed as:

$$U_i = \Pi_{j=1}^{n} x_{ij} \qquad [1.8]$$

In this formula, m and n are the numbers of alternatives and criteria, respectively, and x^{ij} is the performance of ith alternative on the jth objective. The overall utilities are obtained by the multiplication of different units of measurement and become dimensionless. If the decision-makers want to combine the minimization problem with the maximization of other criteria, then [1.8] becomes [1.9]:

$$U_i = \frac{\Pi_{j=1}^{g} x_{ij}}{\Pi_{j=g+1}^{n} x_{ij}} \qquad [1.9]$$

In this formula, g and (n-g) are the number of criteria to be maximized and minimized, respectively. If x_{ij} equals 0, then the withdrawal of that criterion from the decision matrix, because taking 0 value means the absence of a particular criterion in the decision matrix.

1.4. Application of the proposed methodology to the tomato industry

In the tomato processing industry, the quality of the fresh tomato plays a vital role in producing good industrial tomato products to satisfy the final customers. To fulfill this aim, the choice of tomato variety is primordial to ensure the quality of the fresh product.

To ensure the appropriate application of the proposed methodology, an Algerian case study is evaluated in this analysis. An Algerian tomato processing industry needs to select the best tomato varieties in order to cultivate them to ensure the desired objectives, respecting a certain number of criteria.

As mentioned before, the company is located in Blida city in Algeria; it aims to select the best tomato variety to plant in its own fields. The selection of the most appropriate tomato variety is based on the climate condition and the soil characteristics.

In this work, the selection of the most appropriate tomato variety takes into account a certain number of criteria according to the real case study. This contribution is applied to the tomato processing industry in Algeria. In that sense, three MCDMs have been applied to achieve the target (Delphi, AHP and MOORA).

The Delphi method is implemented to determine the criteria, in order to evaluate the alternatives according to the nature of the case study handled. The AHP is applied to identify the weights of the criteria and select the most appropriate variety. Finally, the MOORA approach is used to compare the results obtained using MOORA and AHP methodologies.

1.4.1. *Application of the Delphi method to the case study*

After several visits to tomato processing industries, considerable discussions with the agriculture professionals and analyzing the need of managers to pick the most appropriate tomato variety, it will be clear for us to determine the criteria that must be used in our case study. Afterward, a panel of experts containing seven members (four industry managers and three experts from the academy from the agriculture department and manufacturing and engineering department) participated in that research in order to accurately identify the specific criteria to be properly used for carefully assessing the farms. Given this, the specific criteria considered in this case study are correctly based on agricultural experts' and industry managers' distinct preferences.

Following the Delphi method presented previously, the criteria used in our case study are as follows:

– **Resistance to pest:** excessive levels of temperature and humidity can favor the development of pests and create a lot of diseases in the fruit of tomatoes, which negatively influence the crop. Therefore, the variety of tomatoes that is more resistant to pests is favorable.

– **Productivity:** the objective of farmers is to receive the maximum crops in a limited area. Each variety has its own productivity rate which can produce the difference between different tomato varieties.

– **Cultural flexibility:** the variety of tomato that would be planted must be adapted to soil features, which cause a significant problem of adaptation. The tomato that can adapt with diverse climates and characteristics of soil is more favorable to be picked.

– **Shape and size:** the shape and the size of the fruit have no relevant importance in the tomato processing industry, while the fresh product will be crushed to fabricate tomato paste.

– **Brix:** it is considered one of the most relevant quality indicators in the agri-food supply chain (°Bx). It is a measurement unit of the sugar concentration in fresh tomato and tomato concentrate.

– **Color:** quality tomatoes are firm and uniform in color. The color of the tomato varies, depending on its maturation; it changes from green (beginning of maturation) to red (pick at maturation). When tomatoes are to be used for production, the fruit should be harvested when it is red and fully ripe.

– **Taste:** the taste has a great impact on the sale of the final product. Good tomato variety can have a positive impact on the taste of the manufacturing product to provide it with a sales advantage over other competing manufacturers.

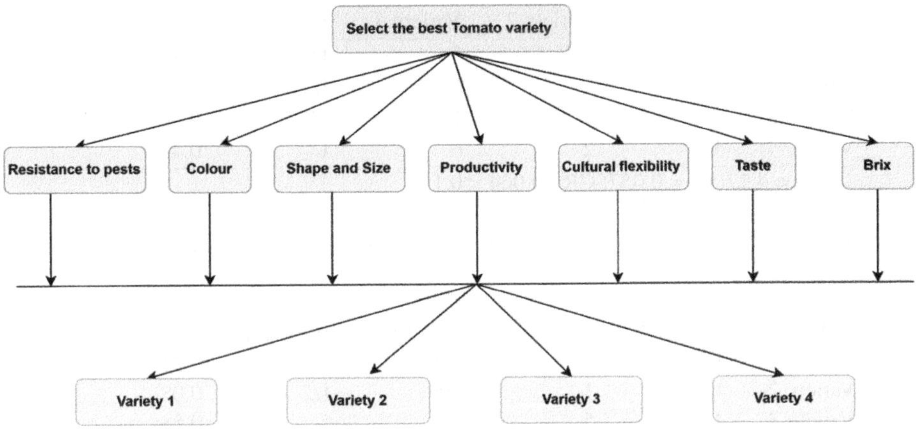

Figure 1.1. *Proposed hierarchical model for tomato variety choice. For a color version of this figure, see www.iste.co.uk/koubaa/optimization.zip*

1.4.2. *Application of AHP methodology to the case study*

As mentioned above, the Delphi methodology is used to identify the criteria. Now, the first step is to construct a hierarchy according to the case study handled as shown in Figure 1.1. The second step is to calculate a pairwise comparison of criteria in order to specify the preference of each criterion based on the opinions of experts. In addition, the alternatives must be evaluated across all criteria, respecting the real case study. Finally, after calculations, each criterion has its weight.

The first pairwise comparison matrix must be calculated, taking the goal into account. The evaluation matrix is presented in Table 1.3. After calculating the criteria weights, the varieties (alternatives) of tomatoes should also be evaluated for each criterion, as shown in Tables 1.4, 1.5, 1.6, 1.7, 1.8, 1.9 and 1.10.

Goal	R. pests	Color	Shape	Productivity	Cultural F.	Taste	Brix	**Weights**
R. pests	1	0.5	5	0.142	0.250	0.333	0.142	0.050
Color	2	1	4	0.142	0.166	0.5	0.142	0.055
Shape	0.2	0.25	1	0.142	0.142	0.333	0.142	0.025
Productivity	7	7	7	1	2	5	0.500	0.248
Cultural F.	4	6	7	0.5	1	5	0.200	0.172
Taste	3	2	3	0.2	0.2	1	0.142	0.068
Brix	7	7	7	2	5	7	1.000	0.378

λ_{max}= 7.805; CI = 0.134; CR= 0.099 <0.1

Table 1.3. *Comparative matrix with respect to the final goal*

R. pests	Variety 1	Variety 2	Variety 3	Variety 4	W
Variety 1	1	1/5	3	1/5	0.11
Variety 2	5	1	6	1/2	0.34
Variety 3	1/3	1/6	1	1/6	0.06
Variety 4	5	2	6	1	0.49
λ_{max} = 4.17; CI = 0.057; CR= 0.064 <0.1					

Table 1.4. *Pairwise comparison matrix for the criterion of resistance to pests*

Color	Variety 1	Variety 2	Variety 3	Variety 4	W
Variety 1	1	3	1/4	2	0.21
Variety 2	1/3	1	1/5	1/2	0.08
Variety 3	4	5	1	5	0.58
Variety 4	1/2	2	1/5	1	0.13
λ_{max} = 4.098; CI = 0.032; CR= 0.036 <0.1					

Table 1.5. *Pairwise comparison matrix for the criterion of color*

Shape	Variety 1	Variety 2	Variety 3	Variety 4	W
Variety 1	1	1/5	1/4	1/6	0.06
Variety 2	5	1	4	1/2	0.32
Variety 3	4	1/4	1	1/4	0.14
Variety 4	6	2	4	1	0.48
λ_{max} = 4.200; CI = 0.066; CR= 0.075 <0.1					

Table 1.6. *Pairwise comparison matrix for the criterion of shape and size*

Productivity	Variety 1	Variety 2	Variety 3	Variety 4	W
Variety 1	1	1/6	1/2	1/6	0.06
Variety 2	6	1	6	3	0.56
Variety 3	2	1/6	1	1/4	0.09
Variety 4	6	1/3	4	1	0.29
λ_{max} = 4.144; CI = 0.048; CR= 0.053 <0.1					

Table 1.7. *Pairwise comparison matrix for the criterion of productivity*

1.4.3. *Application of the MOORA and MULTIMOORA methodologies*

To begin, we construct the decision matrix. The decision matrix indicates the performance of different alternatives with respect to the various criteria. We also add the beneficial criteria and non-beneficial criteria as shown in Table 1.12.

Cultural F.	Variety 1	Variety 2	Variety 3	Variety 4	W
Variety 1	1	1/6	1/2	1/6	0.06
Variety 2	6	1	6	3	0.56
Variety 3	2	1/6	1	1/4	0.09
Variety 4	6	1/3	4	1	0.29

λ_{max} = 4.144 ; CI = 0.048; CR= 0.053 <0.1

Table 1.8. *Pairwise comparison matrix for the criterion of cultural flexibility*

Taste	Variety 1	Variety 2	Variety 3	Variety 4	W
Variety 1	1	1/6	1/2	1/2	0.07
Variety 2	6	1	5	1	0.43
Variety 3	2	1/5	1	1/4	0.10
Variety 4	5	1	4	1	0.40

λ_{max} = 4.035; CI = 0.011; CR= 0.013 <0.1

Table 1.9. *Pairwise comparison matrix for the criterion of taste*

Brix	Variety 1	Variety 2	Variety 3	Variety 4	W
Variety 1	1	1/6	1/2	1/2	0.07
Variety 2	6	1	5	1	0.43
Variety 3	2	1/5	1	1/4	0.10
Variety 4	5	1	4	1	0.40

λ_{max} = 4.154; CI = 0.051 ; CR= 0.057 <0.1

Table 1.10. *Pairwise comparison matrix for the criterion of Brix*

	R. pests	Color	Shape	Prod.	Cultural F.	Taste	Brix	Evaluation	**Ranking**
Variety 1	0.0056	0.0114	0.0015	0.0154	0.0107	0.0047	0.0228	0.0721	4
Variety 2	0.0190	0.0045	0.0083	0.1369	0.0950	0.0299	0.1131	0.4066	2
Variety 3	0.0015	0.0321	0.0036	0.0234	0.0163	0.0073	0.0365	0.1206	3
Variety 4	0.1210	0.0070	0.0122	0.0730	0.0507	0.0270	0.2064	0.4973	1

Table 1.11. *The result of the final evaluation of alternatives*

The second step is to normalize the decision matrix in order to obtain a matrix of a dimensionless number between [0,1], as shown in Table 1.13.

The third and final step is to calculate the sum of the normalized performances for beneficial criteria and subtract for non-beneficial criteria. The results are shown in Table 1.14.

The fourth and final step is to calculate the sum of the normalized performances for beneficial criteria and subtract for non-beneficial criteria. The results are shown in Table 1.13. The best alternative has the highest-ranking value, while the worst alternative has the lowest-ranking value.

For the MULTIMOORA method, we calculate the product of the beneficial criteria divided by the product of the non-beneficial criteria of the normalized decision matrix (Table 1.12). We obtain the results of Table 1.15. The best alternative has the highest-ranking value, while the worst alternative has the lowest-ranking value.

	Min	Max	Max	Max	Max	Max	Max
X^*_{ij}	R. pests	Color	Shape	Prod.	Cultural F.	Taste	Brix
Variety 1	0.0055	0.0114	0.0015	0.0154	0.0106	0.0046	0.0227
Variety 2	0.0189	0.0045	0.0083	0.1368	0.0949	0.0299	0.1130
Variety 3	0.0014	0.0321	0.0035	0.0234	0.0162	0.0073	0.0364
Variety 4	0.1210	0.0069	0.0122	0.0730	0.0506	0.0270	0.2063

Table 1.12. *Normalized decision matrix with each maximized and minimized criteria*

	Min	Max	Max	Max	Max	Max	Max
X^*_{ij}	R. pests	Color	Shape	Prod.	Cultural F.	Taste	Brix
Variety 1	0.0027	0.0056	0.0007	0.0075	0.0052	0.0023	0.0110
Variety 2	0.0093	0.0022	0.0041	0.0669	0.0467	0.0148	0.0546
Variety 3	0.0007	0.0159	0.0017	0.0114	0.0080	0.0036	0.0176
Variety 4	0.0597	0.0034	0.0061	0.0357	0.0249	0.0134	0.0997

Table 1.13. *Normalized performance of alternatives regarding each criterion*

	Evaluation	Ranking
Variety 1	0.0298	4
Variety 2	0.1802	1
Variety 3	0.0577	3
Variety 4	0.1237	2

Table 1.14. *Final results with the application of the MOORA approach*

	Evaluation	Ranking
Variety 1	5.40989E-11	4
Variety 2	8.67102E-08	1
Variety 3	7.94004E-09	3
Variety 4	1.45301E-08	2

Table 1.15. *Final results with the application of the MULTIMOORA approach*

1.5. Discussion

Selection of tomato varieties for planting is one of the most complex problems in the agri-food supply chain due to many hybrid varieties developed annually in laboratories. The managers of companies or farmers always have the conflict problem of choosing the best variety to pick in order to increase the profit. However, the proposed Delphi, AHP, MOORA and MULTIMOORA methodologies can help them to properly choose the best variety among proposed alternatives.

The results obtained using the AHP outputs for the numerical values are shown in Table 1.13. The evaluation of farm selection was realized, and the varieties can be ranked as variety 4, variety 2, variety 3 and variety 1 from the best to the worst according to the AHP numerical evaluation values (high value is more preferable).

Moreover, the application of the MOORA and MULTIMOORA approaches (Tables 1.12 and 1.13) gave us different results compared to the Delphi–AHP methodology. The varieties selected by the MOORA–MULTIMOORA approach are variety 2, variety 4, variety 3 and variety 1, as shown in Tables 1.14 and 1.15.

The difference between the results generated by the different approaches taken into account in this work can be explained by the process of quantifying the data established by experts. Hence, the AHP method uses qualitative values done by experts and verified by CR, the reason why the AHP demonstrates this ability to yield consistent results. However, the MOORA and MULTIMOORA approaches use quantitative values. Also, in our study, we manipulated the qualitative values given by experts and respected the CR as in AHP.

The decision-makers can carefully select one of these methods according to their level of necessary expertise. When the experts undoubtedly possess significant knowledge in the chosen field so the values delivered by them are qualitative, in this situation, the Delphi–AHP methodology must be typically chosen. However, if the professionals have less considerable knowledge of the domain and/or the domain is more technical, the MOORA–MULTIMOORA approaches are preferable if the data provided by experts are mostly quantitative and not qualitative.

Regardless of the choice of the decision-maker, the approach sufficiently taken into account should ideally fit the aspects of the specific problem and its reliable data. In our specific case, the AHP method is the best to choose, because it has the ability to take all the expertise information into account by respecting the CR.

1.6. Conclusion

This study presents a new contribution using Delphi, AHP, MOORA and MULTIMOORA approaches for dealing with suppliers of the best quality tomato for planting. The framework presented in the study is applied to the real case study to illustrate the use of these methods for the TVSP. Given this, the first methodology is a combination of Delphi and AHP methods. The Delphi method has been applied to identify the criteria to be used for evaluation; the AHP approach has been used to determine the criteria weights and make the final decision. The second method is a combination of the MOORA and MUTIMOORA approaches. We have compared these two methods to choose the best tomato variety for planting.

Finally, some relevant characteristics should be mentioned regarding this contribution. It is the first study that addressed the variety selection problem in the tomato processing industry in Algeria. Furthermore, the study can be used as a benchmark for other tomato paste companies or any other company with similar characteristics to this kind of supply chain management.

1.7. Acknowledgment

The authors would like to thank the tomato processing company for its collaboration and the Ministry of Agriculture of Algeria for all information.

1.8. References

Abreu, I. and Mesias, F. (2020). The assessment of rural development: Identification of an applicable set of indicators through a Delphi approach. *Journal of Rural Studies*, 80, 578–585.

Aramyan, L.H., Lansink, A.G.O., Van Der Vorst, J.G., Van Kooten, O. (2007). Performance measurement in agri-food supply chains: A case study. *Supply Chain Management: An International Journal*, 12(4), 304–315.

Brauers, W.K. and Zavadskas, E.K. (2006). The MOORA method and its application to privatization in a transition economy. *Control and Cybernetics*, 35, 445–469.

Bui, T.D., Tsai, F.M., Tseng, M.-L., Ali, M.H. (2020). Identifying sustainable solid waste management barriers in practice using the fuzzy Delphi method. *Resources, Conservation and Recycling*, 154, 104625.

Dožić, S., Lutovac, T., Kalić, M. (2018). Fuzzy AHP approach to passenger aircraft type selection. *Journal of Air Transport Management*, 68, 165–175.

Emovon, I., Norman, R.A., Murphy, A.J. (2018). Hybrid MCDM based methodology for selecting the optimum maintenance strategy for ship machinery systems. *Journal of Intelligent Manufacturing*, 29(3), 519–531.

Giamalaki, M. and Tsoutsos, T. (2019). Sustainable siting of solar power installations in mediterranean using a GIS/AHP approach. *Renewable Energy*, 141, 64–75.

Gupta, T.K., Mangal, D., Srivastava, V.S., Gupta, V.P., Patel, R.V. (2022). A frame work for ranking the factors affecting customer service quality by MOORA approach. *Recent Trends in Industrial and Production Engineering*. Springer, 197–215.

Harputlugil, T., Prins, M., Gültekin, A.T., Topçu, Y.I. (2011). Conceptual framework for potential implementations of multi criteria decision making (MCDM) methods for design quality assessment. *Management and Innovation for a Sustainable Built Environment*, 20–23 June, Amsterdam.

Humphrey-Murto, S. and de Wit, M. (2019). The Delphi method—more research please. *Journal of Clinical Epidemiology*, 106, 136–139.

Joshi, R., Banwet, D., Shankar, R. (2011). A Delphi-AHP-topsis based benchmarking framework for performance improvement of a cold chain. *Expert Systems with Applications*, 38(8), 10170–10182.

Kamali, M., Hosseinnia Kani, S.M., Asghar, A. (2015). Application of Delphi-AHP approach for criteria prioritization of large extractive industrial units site selection. *IMC 2015 International Management Conference*, 26 June, 238.

Kim, M., Jang, Y.-C., Lee, S. (2013). Application of Delphi-AHP methods to select the priorities of WEEE for recycling in a waste management decision-making tool. *Journal of Environmental Management*, 128, 941–948.

Kivak, T. (2014). Optimization of surface roughness and flank wear using the Taguchi method in milling of Hadfield steel with PVD and CVD coated inserts. *Measurement*, 50, 19–28.

Kuram, E. and Ozcelik, B. (2013). Multi-objective optimization using Taguchi based grey relational analysis for micro-milling of Al 7075 material with ball nose end mill. *Measurement*, 46(6), 1849–1864.

Maiyar, L.M., Ramanujam, R., Venkatesan, K., Jerald, J. (2013). Optimization of machining parameters for end milling of Inconel 718 super alloy using Taguchi based grey relational analysis. *Procedia Engineering*, 64, 1276–1282.

Mukeshimana, M.C., Zhao, Z.-Y., Ahmad, M., Irfan, M. (2021). Analysis on barriers to biogas dissemination in Rwanda: AHP approach. *Renewable Energy*, 163, 1127–1137.

Ni-Di, Z. and Ming-Xian, L. (2010). The choice and evaluation of agri-food supplier based on AHP. *2010 2nd IEEE International Conference on Information Management and Engineering*, 484–489.

Oguztimur, S. (2011). Why fuzzy analytic hierarchy process approach for transport problems? *ERSA Conference Papers*, ersa11p438.

Patnaik, P.K., Swain, P.T.R., Mishra, S.K., Purohit, A., Biswas, S. (2020). Composite material selection for structural applications based on AHP-MOORA approach. *Materials Today: Proceedings*, 33, 5659–5663.

Saaty, T.L. (2008). Decision making with the analytic hierarchy process. *International Journal of Services Sciences*, 1(1), 83–98.

Shihab, S.K. and Mubarak, E.M.M. (2017). Multi-objective optimization of wire EDM parameters by applying MOORA technique. *Iraqi Journal of Mechanical and Material Engineering*, 17(2), 354–364.

Wang, W.-P. (2010). A fuzzy linguistic computing approach to supplier evaluation. *Applied Mathematical Modelling*, 34(10), 3130–3141.

Wentholt, M.T., Fischer, A.R., Rowe, G., Marvin, H.J., Frewer, L.J. (2010). Effective identification and management of emerging food risks: Results of an international Delphi survey. *Food Control*, 21(12), 1731–1738.

2

The Effects and Challenges of Climate Change on the Food–Energy–Water Nexus – Poland as a Case Study

The impact of climate change on agriculture has become an issue of greatest concern for researchers, governments and policymakers. Changing climate is a threat to crops, and several studies dealing with the effect of climate change on food production unanimously concluded that negative impacts have already been experienced by food producers worldwide, while a more severe influence is expected to occur in the future. According to the latest predictions for Central Europe, more rainfall is expected in winter, while less rainfall in summer, along with an increased drought risk. Among Eastern European countries, Poland is one of the countries with the smallest water resources per capita and, consequently, water shortage is expected to have far-reaching consequences. In the last few years, Poland has experienced meteorological droughts, which have contributed to water shortages across most of its territory, and reduced flow in rivers and streams with potential to increase the concentration of harmful pollutants. In this work, an analysis of the current trends in climate change and its effects for Central Europe is undertaken using Poland as a case study because of its vulnerability to water shortages. The main challenges and opportunities for Poland related with the Food–Energy–Water nexus are discussed, along with the identification of knowledge gaps and direction for future research that may lead to sustainable, cost-effective, feasible solutions applicable not only to Poland but also to other countries in the region facing similar challenges.

Chapter written by Piotr F. BOROWSKI, Erick R. BANDALA, Aleksandra LUBAŃSKA and Eric STEIN.

2.1. Introduction

Reliable food production depends on the sustained supply of energy and water resources, which highlights the interdependency of the three factors better known as the Food–Energy–Water (FEW) nexus (Berardy and Chester 2017). The increased demand on these key natural resources over the last few decades has led to significant sustainability and environmental challenges, leaving billions of people lacking access to food, energy and/or water (Van Vuuren et al. 2019). In the FEW nexus concept, food, energy and water are intimately related and any change on any of them will impact the other two elements. For example, the total energy demand in the water sector nowadays exceeds 800 TW/h and it is expected to increase ≥ 80% by 2040 (Brouwer et al. 2018), which is expected to translate into increased food prices. Climate change affects the sustainability and security of food, energy and water by modifying weather patterns and cycles, resulting in longer and more severe droughts, as well as more frequent and/or intense extreme meteorological events, wildfires or floods that threaten food production, energy infrastructure and/or water quality (Duran-Encalada et al. 2017). It also alters regional-scale climate patterns in ways that directly affect the FEW nexus and demand further analysis. For example, Central Europe is expected to face a greater incidence of climate extremes resulting from increased radiative force in the region, but the details of this regional variability remain controversial (Topal et al. 2020). The direct effects of climate extremes are most obvious in food production. For example, in 2018, Central Europe faced simultaneous extreme drought and temperature conditions during the crop growing cycle, which eventually resulted in above-normal yield losses (Beillouin et al. 2020). Other significant indirect effects include deterioration of groundwater quality due to decreased aquifer recharge, combined with an accumulation of contaminants of anthropogenic origin, and combined effect of climate and land-use changes (Nistor 2020). After the drastic changes introduced after 1989 in Central Europe, countries in the region have moved at different speeds toward free market economies, as well as in food and energy production or water management (Brondy et al. 2020). In many regards, these changes occurred at a slower pace in Poland than in other countries in the region, leading to struggles in the country's energy- and environment-related problems and exacerbated by the effects of climate change on water availability and food production (Kaczmarczyk et al. 2020). This study analyzed the current trends in climate change and their effects on Central Europe. Poland was used as a case study because of its vulnerability to water shortages, energy production and food security. The main challenges and opportunities for Poland related to the FEW nexus are discussed, and knowledge gaps as well as directions for future research that may lead to sustainable, cost-effective, feasible solutions also applicable to other countries in the region facing similar challenges.

The research was carried out using quantitative and qualitative methods. The qualitative methods used observation techniques and desk research techniques. The authors made observations on the relationship between FEW and climate change. Desk research was based on the analysis of available sources in scientific journals, reports and previously published articles on the subject. Quantitative methods were used in food production issues, the conversion rate (CR) was calculated as well as the method used for ecological/chemical improvement in water-related issues. The methodology used in the study is shown in Figure 2.1.

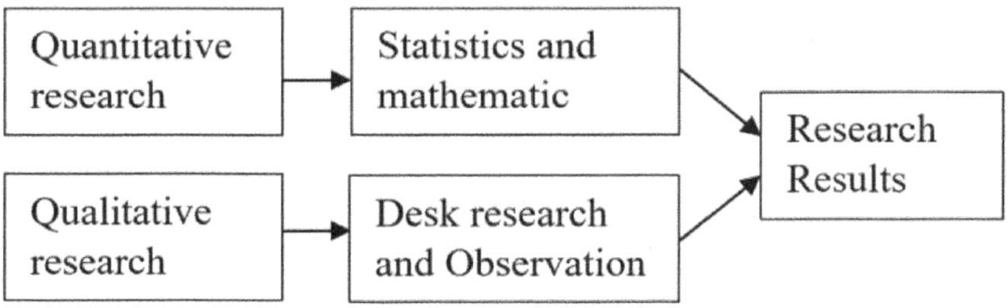

Figure 2.1. *The methodology scheme used in the research*

2.2. Effect of climate change scenarios on Poland's FEW nexus

Europe is increasingly exposed to the effects of climate change. For several years, climate change has increased average temperatures and changed the amounts and types of precipitation (Borowski 2020a). These effects pose a major challenge to agriculture and the environment (Kelemen et al. 2009). In comparison to the 1850–1900 average, the global increase in surface temperature at the end of the 21st century is predicted to range from 1.5°C to 4°C, depending on the scenario adopted (Masson-Delmotte et al. 2018).

The actions of individual Central European countries, including Poland, undertaken in response to climate challenges will play a key role in mitigating the negative effects associated with climate change. The average annual air temperature in the 1951–2008 period increased in Poland within the range from 1°C to 2°C. Although major changes are expected during the winter, average seasonal temperatures are also expected to increase throughout Poland in all seasons. The greatest warming will occur in winter and will exceed 0.6°C in almost the entire country. All emission scenarios for the winter period consistently predict a nationwide increase in temperature from 0.08°C to 0.72°C calculated in relation to the average temperatures in the 1971–1990 period. The scenarios for precipitation show a zone of increasing rainfall in the east, but the dominant feature is to decrease in rainfall covering most of the country approximately 7–8%, and even over 10% locally (Wibig and Jakusik 2012).

2.2.1. *Poland's food production scenarios*

Climate change has negatively affected crop production in Poland. The drought period has extended mainly in the period of emergence (April–May) followed by violent, intense, long-lasting rainfall that causes crop flooding, mainly during harvest season (Kundzewicz and Matczak 2012). Therefore, preventing the effects of climate change has become a topic of concern, particularly for small vegetable producers located near big urban communities (Soika et al. 2020). Current measures involve production planning over the following years and including moisture- and drought-resistant varieties in the seed and seedling selection (Ratajczak et al. 2019). Although using resistant crop varieties does not directly increase production costs, it affects the quality of the final product, particularly for retail or wholesale vegetables.

Other significant challenges are determining the distribution of certain vegetables in specific fields for a given season and the appropriate crop rotation. Due to the effects of climate change, vegetable producers have changed their strategies for selecting fields for a specific plant/vegetable for several years now. In the past, for example, the selection of crop sites (soil type) was made based on the plant's optimal soil requirements. Nowadays, the positions are differentiated to minimize losses due to downpours, flooding and droughts. Both heavy and light soils are selected for each vegetable, and unfavorable climate change effects result in lower yields with the consequent drop in production yields in recent years, as shown in Table 2.1.

Product (vegetable)	**Production yield [CR] (ton/ha)**		
	2018	**2019**	**2020***
Onion	35 [0.93]	35 [0.92]	30 [0.86]
Potato	46 [0.92]	48 [0.90]	45 [0.85]
Carrots	82 [0.94]	90 [0.96]	75 [0.88]
Pumpkin	33 [0.86]	32 [0.86]	24 [0.82]

* 2020 is rated with extremely unfavorable climatic conditions. Sources: own elaboration.

Table 2.1. *Production yields for selected vegetables in Poland, 2018–2020*

Repeated droughts and/or heavy rains have had a negative impact not only on the amount of the vegetable harvest, but also on their quality. For example, long-lasting rains during the onion harvest greatly reduced the quality of the product, which affects the final selling price. To better evaluate the effects of climate change, the CR for individual products is used. The CR expresses how large a percentage of the product that goes to the production line is suitable for sale as a final product. The higher this coefficient, the better for the manufacturer. The CR value depends mainly on the quality of the raw material and the storage process. During the storage process, vegetable disease processes may intensify, reducing the profitability of the final product and causing a greater number of complaints by the buyers. In 2020, the CR value was at its lowest (Table 2.1) because of too much moisture and flooding (Farias et al. 2017).

2.2.2. *Poland's energy usage scenarios*

European Union (EU) countries have adopted solutions in the fields of energy and climate policy, including efforts to reduce greenhouse gas emissions, improve energy efficiency and pursue renewable energy sources (RES) by 2030. Therefore, the EU has made significant and noticeable progress in implementing its plans for a single, internal electricity market in recent years (European Union 2020), including currently undergoing major changes in the power system (Newbery et al. 2018; Korzarcanin et al. 2019).

Expansion of energy transmission networks and an improved management framework are planned in the coming years, as well as updates to the CO_2 emission standards for new vehicles. The new framework not only concerns management at the economy and national policy levels, but also enterprises and their adaptation to technological, ecological and social environments (Borowski 2020b). The significant efforts required to achieve climate neutrality entail fundamental transformation in all sectors of the economy and mobilization of a wide range of resources, covering both technological and financial dimensions, as well as consumer preferences and choices (World Energy Council 2016).

In Poland, energy policy has been traditionally dominated by stakeholders of high-emission industries, primarily mining and coal-based energy. Each economic sector requires a specific energy service to achieve its projected activity level each year. The energy service is a logistic chain (Borowski 2020c), and it involves optimizing both material resources and equipment usage, as well as using energy-intensive equipment under various constraints. The transmission of energy from the power plant to the recipient is possible through a long network of power lines and stations, and associated with large energy losses. The basic way to reduce these losses is to increase the voltage of power transmission lines. Depending on the distance over which the energy is to be transmitted, different voltage values are applied. Raising the voltage for transmission purposes, and then lowering it to the level at which it is possible to use electrical appliances built for 220/230 or 380/400 V, requires using high-voltage-system power stations, many high-voltage distribution stations and numerous transformer stations to convert medium voltage (distribution) into those commonly used in receiving installations (230/400 V). This is one of the investments that will be needed in the coming years, an infrastructure upgrade based on the traditional model of the electricity market (Jeszke 2019).

Electricity consumption and final energy demand are steadily increasing in all sectors, and they are projected to reach 25% above the 2015 level by 2050. However, not all factors influencing electricity consumption are predictable. For example, the Covid-19 pandemic reduced energy demand in Poland. Electricity consumption was 6.9% lower than during the same period in 2019 and 8.1% lower compared with the same period in 2018. The average daily energy consumption was 6.5% lower (Czosnyka et al. 2020).

2.2.3. *Poland's water security scenarios*

In terms of available freshwater resources in Central European countries, Poland has the second lowest amount of renewable water resources per 1,000 inhabitants (1,600 m^3 per capita), following only Czechia, which has 1,500 m^3 per capita. A country's freshwater availability is significantly determined by climate conditions, transboundary water flows and total population. Water-rich European countries have resources as high as 20,000 m^3 per capita or higher, whereas countries below the 1,700 m^3 per capita threshold are experiencing water stress. Poland is one country with just a slight dependence on the external inflow (12.7%) (WWP 2017). The main challenges for water management in Poland is to satisfy the increasing demand of fresh water while protecting aquatic ecosystems and ensuring that scarce and variable water resources remain available to avoid water insecurity. Agricultural water use in Poland is 10% of total water consumption, significantly lower than the regional and global annual agricultural water consumption (40% and 70% for the European and global averages, respectively); the other two major water uses reported are 19% for households and 70% for industrial purposes (Kubiak-Wojcicka and Machula 2020). Nevertheless, the greatest increase in agricultural water use in Europe is expected in coming years because of climate change (Dolganova et al. 2019). To avoid water-related food insecurity, Poland needs to start considering other approaches to water management so that the expected increase in water for agriculture will not limit the productivity of agricultural land (U.S. Department of State 2020).

The quality of surface and groundwater in Poland is highly sensitive to anthropogenic effects and exhibits significant spatial and temporal variability. Most conducted studies pertain to surface water, whereas significantly fewer studies have been carried out for groundwater, highlighting a lack of information regarding water quality and/or the impact of environmental stress factors.

2.3. Adaptation measures to the effects of climate change for the FEW nexus in Poland

2.3.1. *Food production adaptation*

Adaptations in response to climate change to support food production involve actions taken to reduce risk and vulnerability and increase resilience by adjusting practices, processes and capital in response to observed or actual climate change effects (Adger et al. 2007; Howden et al. 2007; Porter et al. 2014). Adaptation can occur at multiple system levels: physical, ecological and human (Adger et al. 2007; Howden et al. 2007), and its magnitude can further be classified as incremental, systemic and transformative (Vogel and Meyer 2018). Adaptation is context specific, output specific, and it is a function of geography, local conditions, agricultural practices and methods, and socioeconomic, political and organizational factors (Belliveau et al. 2006; Vogel and Meyer 2018). Several options have been proposed to adapt traditional farming methods in response to climate change. Vogel and Meyer (2018) articulated a comprehensive list of adaptation types, ranging from incremental to systemic (Figure 2.2).

As suggested in Figure 2.2, from a horticultural perspective, implementing changes to the plants themselves is one option, including breeding more adaptive plants to changing environments, resistant to disease and pests, able to enhance nutrient content (Vogel and Meyer 2018) and/or appropriate varieties for specific environments (Schneider and Asch 2019). Other options include controlling the nursery environment, and then transplanting the crops to the field (Seinn and Ahmand 2015). A different set of adaptions include changes in farm management methods, optimizing row spacing and density (Vogel and Meyer 2018), modifying alternative nutrient management and crop rotation, and using advanced data collection and predictive models to optimize nutrient applications (Schneider and Asch 2019).

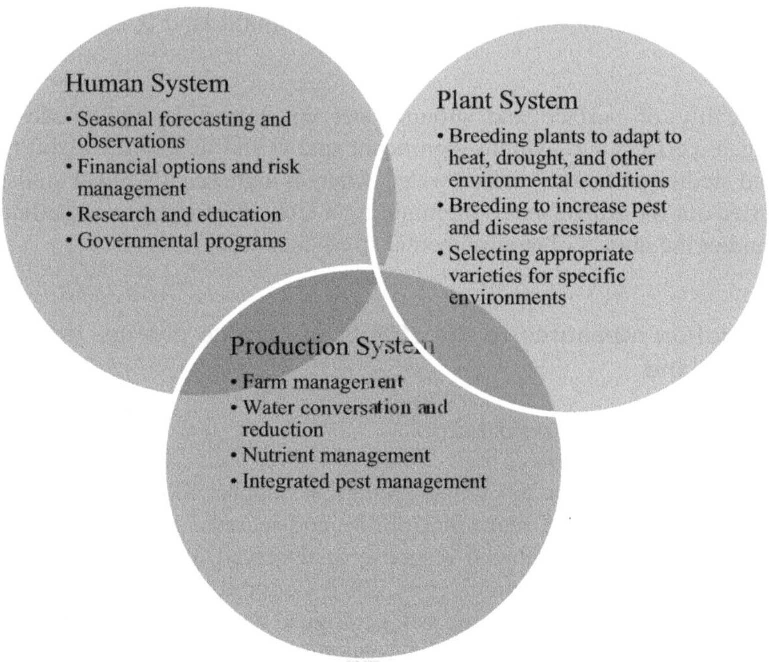

Figure 2.2. *Adaptation options for traditional agriculture (adapted from Vogel and Meyer (2018)). For a color version of this figure, see www.iste.co.uk/koubaa/optimization.zip*

Food production in urban areas will be an increasingly important means to adapt to the high consumption of resources around cities (Gondhalekar and Ramsauer 2017). Applying the concepts of the FEW nexus originally developed for large-scale food production to cities is appropriate if the relatively small-scale size of current production units and other unique characteristics of urban farming is considered (Caputo et al. 2021). Growing food in urban areas is not new (Stein 2021), and it is also referred to as urban horticulture and fosters cultivation of food and ornamental plants (Khan et al. 2020), with potential to meet 18–100% of local vegetable demand using only 7–10% of the existing space (Caputo et al. 2021). It offers a means to simultaneously mitigate effects of climate change and produce food, thereby taking stress from traditional forms of agriculture. To maximize the contribution of urban agriculture to climate change adaption, greater levels of planning and professional development for farmers and implementing key information technologies will be needed.

The implementation of indoor vertical farming in Poland's economy is expected to produce sustainable benefits, such as reducing pesticide use, mitigating excess fertilizer runoff and effectively using non-arable land (Stein 2021). Indoor farming methods can mitigate the negative impacts of agricultural food production on the aquatic food production capabilities of freshwater bodies. Shifts to indoor farming can also mitigate effects of traditional agriculture on soil degradation (Ramzan et al. 2020) and reduce CO_2 emissions (Stein 2021). Its application, however, requires external sources of lighting to support plant growth year-round. The amount of light required by a plant per day varies by cultivar and location and is expressed as the daily light integral (DLI). The DLI is a measure of the number of active photons (400–700 nm) delivered to a specific location over a 24-hour period. The role of DLI on plant growth has been studied extensively (Wang et al. 2020) and possesses implications for energy use. For example, a plant factory growing lettuces will target 15–20 mol/m^2/d with a photoperiod of 16 h/day operating 365 days per year (Brechner and Both 2013). This DLI can be achieved with high-performance, modern LEDs positioned 0.3–0.4 m from the canopy requiring 79 W/m^2. A large-scale lettuce plant factory (e.g. 4,600 m^2) will consume 400 kW instantaneously and require 2.34 MWh per year. The energy requirement for fruiting plants is two- to three-folds greater (Faust and Logan 2018). One strategy to mitigate this load is incorporating RES into the facility design or buying energy from the grid from sustainable sources. In addition, indoor farms include significant heating, cooling and ventilating costs. Incorporating geothermal energy is another viable alternative to controlling energy costs, even considering that lighting itself will generate significant heat (1 kWh = 3,600 kJ).

2.3.2. *Energy usage adaptation*

Adapting the energy sector in Poland to climate change scenarios can be conceptualized as actions needed in several main areas: (1) energy security; (2) internal energy market; (3) emission reduction; and (4) research science, innovation and competitiveness. The energy security of Poland plays a key role as a comprehensive concept that applies to many different areas combining energy, economic growth and political power. It is characterized by changes and modifications occurring in individual countries, as well as regional and global context (Nyga-Lukaszewska et al. 2020). For the next 30 years, the construction of nuclear power plants supplemented by renewable energy facilities is planned to meet EU energy and climate policy requirements, electricity demand and energy source diversification, and help reduce national greenhouse gases emissions. Nuclear power plants are expected to become the basis for the operation of the Poland power system (Zawalinska et al. 2020). The commissioning of the first unit (capacity 1–1.5 GW) is scheduled for 2033 and another five are planned by 2043 (Szabo and Fabok 2020). Using RES demands efforts to ensuring energy security. As many external factors (windiness, solar radiation or water levels in rivers) are involved, energy shortage is expected to occur (Brondy et al. 2020). Strategies to avoid these disruptions involve using energy surpluses from other periods, effective energy storage and collaboration among users (Lewandowska et al. 2020).

Poland significantly depends on imported crude oil, which requires active policy managing petroleum-derived fuel demand. To achieve this goal, popularization of alternative transport fuels, which has been encouraged by the Polish government in the Act on Electromobility and Alternative Fuels, is fundamental. The Act provides incentives to stimulate electromobility development and other alternative fuels (e.g. Liquefied Natural Gas (LNG) and Compressed Natural Gas (CNG)) in transport (Burchart-Korol et al. 2020). Dependence on crude oil imports will be reduced by increasing the use of alternative fuel-powered cars, expanding charging infrastructure for electric cars and refueling LNG/CNG in the transport sector. The following ambitious goals to expand infrastructure for alternative fuels by 2030 have been adopted: (i) construction of 6,000 points for electric car power charging, (ii) construction of 70 CNG refuel sites and (iii) opening 32 publicly accessible CNG refuel points (Burchart-Korol et al. 2020). By 2050, the Polish energy sector is expected to become climate neutral (Piotrowski et al. 2020).

Poland's internal electricity market aims to ensure proper energy sector functioning, supports EU's decarbonization, and electricity cross-border trade by enabling clean energy transition, ensuring climate legislation by 2030 and honoring the Paris Agreement. The regulatory framework for energy infrastructure will be reviewed to ensure consistency with the European Green Deal's goal of climate neutrality (Montanarella and Panagos 2021).

Carbon emission reduction will be achieved by implementing activities aimed at disseminating the circular economy (CE) intended to reduce environmental impacts of manufactured products, minimizing waste generation and achieving climate neutrality (Borowski 2021). From an energy perspective, the CE focuses on raw material management to reduce waste by converting it into an energy source. Over 80% of energy in Poland is generated from conventional sources and plans are to reduce the use of anthracite and lignite to 60% by 2030, with a further downward trend until 2040. Poland also intends to achieve 21% of RES by 2030 to reduce GHG emissions in non-Emissions Trading System (ETS) sectors (Hasterok et al. 2021).

In 2019, 159 TWh electricity was generated while consumption was 169 TWh, and the difference was balanced with international exchange (Borowski 2020). By 2030, Poland's capacity for cross-border electricity connections is expected to increase and countermeasures for climate change are also planned (Burchart-Korol et al. 2020). Innovations in the energy sector will reduce operating costs and provide improved product and service to better meet market expectations. The use of Industry 4.0 solutions is expected to increase, improving operational efficiency, changing work organization, improving efficiency, and enabling new operating standards and environmental monitoring mechanisms (Borowski 2021). The sector is expected to depart from conventional models that consider energy consumers passively and predictably receiving electricity or heat (Kundzewicz and Matczak 2012; Burchart-Korol et al. 2020). Sector digitization will optimize power grid operations and increase the distribution network efficiency (Borowski 2020b). New technologies will completely change the current energy sector, increasing research and development expenditure in Poland to 1.7% of GDP in 2020 and 2.5% of GDP in 2030 (Kundzewicz and Matczak 2012).

2.3.3. *Water security adaptation*

2.3.1.3. *Water management*

The effort related to water management achieved in Central European countries, including Poland, after implementing EU Water Framework Directive (WFD) are undeniable. Nevertheless, to comply with the WFD objective to achieve good economical and chemical status in all EU water bodies, more work is needed related to the FEW nexus to prevent food, energy and/or water insecurity. Achieving good chemical and ecological indicators of water bodies is a significant challenge as less than 40% of EU surface water bodies reached good ecological status in the first management cycle, and 44% are expected to achieve a good status for the second management cycle (Machac et al. 2020). Table 2.2 shows the registered improvement in ecological and chemical status for surface water bodies in Central European countries (European Environmental Agency 2018); it is worth noting that although Poland showed significant improvement, these numbers are due to the lack of information reported for the first WFD management cycle.

For example, a recent study collected information related to water supply and sanitation infrastructure imbalance in Poland, particularly in rural communities (Piasecki 2020). The impact of this imbalance on water management is significant because domestic sewage is discarded into the soil or a nearby stream, increasing deterioration of the receiving water body and risk for epidemic outbreaks.

The amount of wastewater, either treated or untreated, released into surface water or soil in Poland during 1998–2017 systematically increased. It is worth noting the significant increase in the amount of wastewater discharged over the last two decades. Also, the amount of untreated sewage reduced to none after 2009, and only two major water treatment processes have been used in the last few years. Despite undeniable efforts to improve wastewater treatment, conventional wastewater treatment processes are unable to remove certain types of contaminants (Clurman et al. 2020). Therefore, relying only on these two processes not only presents significant challenges for the country's FEW nexus, but also for its sustainable development. Novel wastewater treatment technologies are needed to prevent pollutants other than conventional contaminants (e.g. viruses and/or emerging, persistent and toxic contaminants) from ending up in water sources. The search for alternative wastewater treatments or adaptations for currently available water treatment technologies that address Poland's specific needs is an interesting direction for future research that was identified in this study.

Country	Ecological status		Chemical status	
	WFD management cycle		WFD management cycle	
	First (%)	Second (%)	First (%)	Second (%)
Austria	42	47	99	0
Czechia	17	19	70	69
Germany	9	8	88	0
Poland	2[1]	30	3[1]	59
Slovakia	63	57	95	98
EU average	39	44	37	31

Estimated by the European Environmental Agency because most values were unknown.

Table 2.2. *Registered improvement in ecological/chemical status in Central Europe countries (adapted from European Environmental Agency (2018))*

In Poland, stormwater has historically been treated as waste and discarded immediately, generating a significant lack of consistency and stability of legal regulations for implementing sustainable stormwater management in the country (Kordana and Slys 2020a), in agreement with other megalopolises in developing countries (Goonetilleke et al. 2017). Some recent studies, however, have identified great potential for stormwater reuse in Poland, but very little information is available (Kordana and Slys 2020a, 2020b). The need for fit-for-purpose treatment of stormwater effluents before agricultural use is evident because of the presence and concentration of heavy metals (Bak et al. 2019). The presence of heavy metals in stormwater runoff significantly affects macroinvertebrates in urban rivers and reservoirs, eliminating the possibility of reuse until proper treatment is carried out (Goldyn et al. 2018). The lack of information on the potential for stormwater catchment, treatment and reuse in Poland is another significant knowledge gap worth exploring because stormwater could replace freshwater sources for crop production, provided there is a proper fit-for-purpose treatment to ensure food security; this may allow conventional water sources to be devoted to other uses such as drinking water or specific manufacturing applications.

One ambitious WFD goal requires planning from an integrated point of view (Satalova and Kenderessy 2017). Poland is moving toward having operational planning/implementation at the river basin scale in the last two decades (Jager et al. 2016). Considering the FEW nexus, public participation is needed for proper water resource management because current and future users share decision-making and technology implementations to enhance water, food and/or energy security. Therefore, sharing information to improve communication, encouraging power delegation and developing interventions are necessary to ensure basin management.

Figure 2.3 shows the major factors involved in integrated watershed management in Poland and other Central European countries (Ptak et al. 2020). All the challenges identified are highly significant knowledge gaps worth exploring and require capacity building programs to address watershed health and function protection, and development/enhancement of local actions that contribute to Poland's self-reliance.

2.3.3.2. *Water quality*

Emerging contaminants (ECs) originate from diverse sources and the typical range in concentration from ng L^{-1} to μg L^{-1} (Rodriguez-Narvaez et al. 2017). They include natural and synthetic chemical compounds that are commonly present in water, which have only recently been recognized as significant water pollutants (European Environmental Agency 2018). In recent years, the use of activated carbon, biochar, and hydrochar has been investigated for the adsorption of ECs with success because of their high porosity and specific surface area (Bandala and Rodriguez-Narvaez 2019). Recent research (Rodriguez-Narvaez et al. 2019) conducting a deep analysis on the use of carbon-based materials to treat water contaminated with ECs highlighted their cost-effectiveness, high potential for use in developing countries and tunable nature that allows them to remove target contaminants (Clurman et al. 2020).

Harmful algal bloom (HAB) events are hard to predict and very quickly attain high densities, with devastating effects on biota and long-term impacts on lake ecology. Although high nutrient loads (e.g. N and P) are closely linked to HABs, the response of algal communities to nutrient load, pH, temperature, dissolved oxygen and several other water quality parameters is unknown, as are the mechanisms for HABs in fresh water (Royer 2020). Algal blooms are best managed at a local level because of the varying circumstances of each event. Local councils and state water authorities should investigate suspected outbreaks and alert potential users to avoid using unsafe water for food production or other uses.

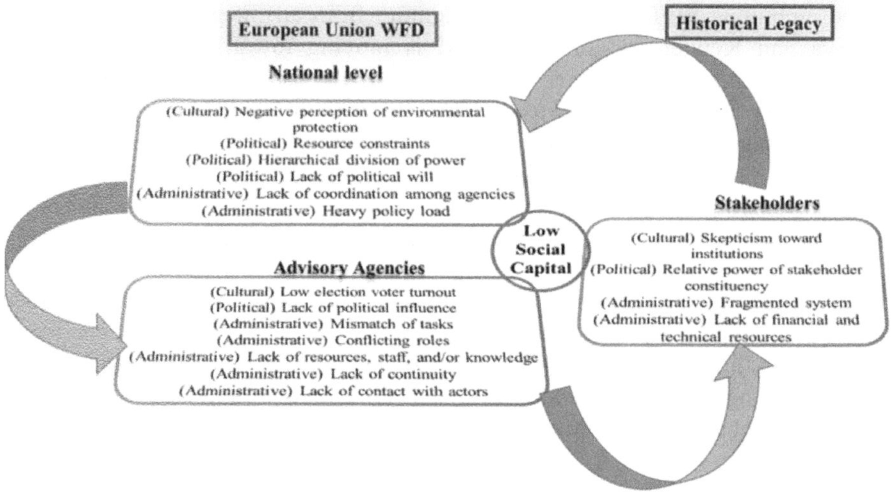

Figure 2.3. *Major factors constraining integrated watershed management in Poland (adapted from Ptak et al. (2020))*

The most effective and plausible method for stymieing algal blooms is to incorporate a comprehensive watershed management plan to minimize the nutrient load entering waterways through nonpoint sources. To implement this plan, information access and a better understanding of the conditions that trigger HABs are needed. The application of state-of-the-art technologies is key to achieving these goals. To be sure, users in the FEW nexus must prevent algal blooms through comprehensive approaches that engage numerous actors in the development and integration of the best practices for their watersheds. Implementing such a broad system can be costly and will require a highly coordinated approach among many actors. Nevertheless, costs can be reduced, and coordination can be achieved using technologies such as wireless communication, water-sensing arrays or the IoT. We believe that developing a cutting-edge application of state-of-the-art technologies to monitor and understand HABs will have many long-term benefits for people in the communities involved in the FEW nexus. Deployment of such technologies in developing countries with significant vulnerability to climate change, such as Poland, will help develop proper adaptation measures to ensure water, food and energy security amid the diverse climate change scenarios.

Microorganisms have been known to infect plants and destroy crops, as well as increase the risk of epidemic outbreaks that threaten human lives (Aurioles-López et al. 2016; Huesca-Espitia et al. 2017). For example, among the different microorganisms, viruses possess a relatively higher resistance to conventional water disinfection processes (Garcia-Gil et al. 2020), which highlights the need for novel, cost-effective methodologies for their inactivation in water before it is used for food production or human consumption (Adelodun et al. 2020; Zhang et al. 2020). Conventional, secondary wastewater treatment processes can remove 2 to 3 log_{10}-cycles of virus content through adsorption to the solid particles of activated sludge (Zhang et al. 2018).

Membrane technologies (e.g. ultrafiltration, nanofiltration and reverse osmosis) are very efficient at removing viruses and/or solid-associated viruses, achieving removal efficiencies ranging from 0.2 log_{10}- to 6.5 log_{10}-cycles (Bhowmick and Dhar 2019), but they only involve mechanical separation and not the inactivation of the pathogens. In this regard, other technologies have been tested for the inactivation of different pathogenic microorganisms in water. It is important to note that different advanced oxidation processes (AOPs) have been tested and identified as having great potential and providing a variety of alternatives for pathogen inactivation. For example, using either solar disinfection, ozonation, wet oxidation or even cold plasma to generate reactive oxygen species (ROS) has shown significant potential for inactivating bacteria and other pathogenic microorganisms (Castillo-Ledezma et al. 2011, 2014; Aurioles-López et al. 2016; del Carmen Huesca-Espitia et al. 2017; Huesca-Espitia et al. 2017). Other methodologies have been reported, including conventional chlorination procedures, the use of alternative disinfectants (Garcia et al. 2019; Rachmadi et al. 2020) and the application of high-temperature CO_2 injection in water for pathogen inactivation (Sanchis et al. 2019). For example, Fenton and Fenton-like processes are effective alternatives for inactivating highly resistant microorganisms (Bandala et al. 2011). Considering the threat posed by microorganisms such as phytopathogens to decimate crops and put human populations at risk of waterborne disease outbreaks, developing cost-effective technologies to minimize the release of pathogens into the environment through wastewater effluents or their reuse in agriculture is a highly desirable scientific task. In developing countries, the available infrastructure may not suffice to ensure water or food security, particularly in rural communities that lack access to resources to properly disinfect water using conventional water treatment processes.

2.4. Conclusions

In this study, the main challenges and opportunities for Poland related to the food–energy–water nexus are discussed and reported, as well as knowledge gaps and future research areas being identified, which may lead to sustainable, cost-effective, feasible solutions for Poland and other countries in the region facing similar challenges. The following are our main findings:

– Europe is increasingly exposed to the effects of climate change. For several years, climate change has gradually increased average temperatures and changed the amounts and types of precipitation. The main impact of these changes will be felt in the long term. However, the effects of the increasing frequency of extreme weather conditions are already noticeable and pose a major challenge to agriculture and the environment.

– Based on the conducted observations, the average annual air temperature in the 1951–2008 period increased in Poland from approximately 1°C in the eastern part of the country to almost 2°C. Temperature changes in Poland for 2011–2030 are predicted to differ depending on the seasons and regions.

– The relationship between the food–energy–water nexus will also influence climate change and ecosystems in Poland.

– The main adaptation strategies for food production were identified as improving traditional farming practices, expanding urban agriculture as well as indoor and vertical farming methods and encouraging the implementation of various "smart farming" methods to increase productivity, maintain low production cost, save water, energy and ensure food safety for the final consumers.

– The five main areas identified as major energy usage adaptation were energy security, internal market, efficiency, emission reduction and research, innovation and competitiveness, considered strategic for Poland's path to self-resilience.

– In the case of water security adaptations, improving water management and improving water quality were found basic to ensure water security, including wastewater management, stormwater collection, treatment and reuse, integrated watershed management, emerging contaminant removal and HAB prevention.

The results of the research conducted allow for the conclusion for the future that the nexus between food, energy and water will play a key role in mitigating the effects of climate change. The possibility of a better, more efficient use of energy in the economy as well as the economic management of water resources and the implementation of sustainable agricultural production are closely related. Proper management of the FEW nexus will allow for the functioning of an environmentally friendly economy.

2.5. List of abbreviations

AOP	Advanced Oxidation Process
CE	Circular Economy
CNG	Compressed Natural Gas
CR	Conversion Rate

DLI	Daily Light Integral
EC	Emerging Contaminants
ETS	Emissions Trading System
FEW	Food–Energy–Water
HAB	Harmful Algal Bloom
LNG	Liquefied Natural Gas
RES	Renewable Energy Sources
WFD	Water Framework Directive

2.6. Acknowledgments

This manuscript is based on work supported in part by the Nevada Water Resources Research Institute (Grant No G16AP00069). The authors are also grateful to Ms. Nicole Damon (DRI) for her editorial review.

2.7. References

Adelodun, B., Odedishemi, F., Gbemisola, R., Olalekan, H., Choi, K. (2020). Snowballing transmission of COVID-19 (SARS-CoV-2) through wastewater: Any sustainable preventive measures to curtail the scourge in low-income countries? *Sci. Total Environ.*, 742, 140680. doi: 10.1016/j.scitotenv.2020.140680.

Adger, W.N., Agrawala, S., Mirza, M.M.Q., Conde, C., O'Brien, K., Pulhin, J., Pulwarty, R., Smit, B., Takahashi, K. (2007). Assessment of adaptation practices, options, constraints and capacity. In *Climate Change 2007: Impacts, Adaptation and Vulnerability*, Parry, M.L., Canziani, O.F., Palutikof, J.P., van de Linden, P.J., Hanson, C.E. (eds). Vulnerability, Cambridge, 717–743.

Aurioles-López, V., Polo-López, M.I., Fernández-Ibañez, P., López-Malo, A., Bandala, E.R. (2016). Effect of iron salt counter ion in dose-response curves for inactivation of *Fusarium solani* in water through solar driven Fenton-like processes. *Phys. Chem. Earth*, 91, 46–52. doi: 10.1016/j.pce.2015.10.006.

Bak, L., Szelag, B., Gorski, J., Gorska, K. (2019). The impact of catchment characteristics and weather conditions on heavy metal concentrations in stormwater – Data mining approach. *Appl. Sci.*, 9, 2210. doi: 10.3390/app9112210.

Bandala, E.R. and Rodriguez-Narvaez, O.M. (2019). On the nature of hydrodynamic cavitation process and its application for the removal of water pollutants. *Air, Soil Water Res.*, 12. doi: 10.1177/1178622119880488.

Bandala, E.R., Perez, R., Velez-Lee, A.E., Sanchez-Salas, J.L., Quiroz, M.A., Mendez-Rojas, M.A. (2011). *Bacillus subtilis* spore inactivation in water using photo-assisted Fenton reaction. *Sustain. Environ. Res.*, 21, 285–290.

Bandala, E.R., González, L., Sanchez-Salas, J.L., Castillo, J.H. (2012). Inactivation of Ascaris eggs in water using sequential solar driven photo-Fenton and free chlorine. *Journal of Water and Health*, 10(1), 20–30.

Beillouin, D., Schauberger, B., Bastos, A., Ciais, P., Makoswki, D. (2020). Impact of extreme weather conditions on European crop production in 2018. *Philos. Trans. B.*, 375, 20190510. doi: 10.1098/rstb.2019.0510.

Belliveau, S., Smit, B., Bradshaw, B. (2006). Multiple exposures and dynamic vulnerability: Evidence from the grape industry in the Okanagan Valley, Canada. *Glob. Environ. Chang.*, 16, 364–378. doi: 10.1016/j.gloenvcha.2006.03.003.

Berardy, A. and Chester, M.V. (2017). Climate change vulnerability in the food, energy, and water nexus: Concerns for agricultural production in Arizona and its urban export supply. *Environ. Res. Lett.*, 12, 035004. doi: 10.1088/1748-9326/aa5e6d.

Bhowmick, G.D. and Dhar, D. (2019). Coronavirus disease 2019 (COVID-19) outbreak: Some serious consequences with urban and rural water cycle. *Clean Water*, 3, 1–8. doi: 10.1038/s41545-020-0079-1.

Borowski, P. (2020a). Nexus between water, energy, food and climate change as challenges facing the modern global European and Polish economy. *AIMS Geosci.*, 6, 397–421. doi: 10.3934/geosci.2020022.

Borowski, P. (2020b). Zonal and nodal models of energy market in European Union. *Energies*, 13, 4182. doi: 10.3390/en13164182.

Borowski, P. (2020c). New technologies and innovative solutions in the development strategies of energy enterprises. *High Tech. Innov. J.*, 1, 39–58. doi: 10.28991/HIJ-2020-01-02-01.

Borowski, P. (2021). Innovation strategy on the example of companies using bamboo. *J. Innov. Entrep.*, 10. doi: 10.1186/s13731-020-00144-2.

Brechner, M. and Both, A.J. (2013). *Hydroponic Lettuce Handbook.* Cornell Controlled Environment Agriculture, Cornell University CEA Program, New York.

Brondy, J., Tutak, M., Saki, S.A. (2020). Forecasting the structure of energy production from renewable energy sources and biofuels in Poland. *Energies*, 13, 2539. doi: 10.3390/en13102539.

Brouwer, F., Avgerinopoulos, G., Fazekas, D., Laspidou, C., Mercure, J.F., Pollitt, H., Ramos, E.P., Howells, M. (2018). Energy modelling and the Nexus concept. *Energy Strateg. Rev.*, 19, 1–6. doi: 10.1016/j.esr.2017.10.005.

Burchart-Korol, D., Gazda-Grzywacz, M., Zarebska, K. (2020). Research and prospects for the development of alternative fuels in the transport sector in Poland: A review. *Energies*, 13, 2988. doi: 10.3390/en13112988.

Caputo, S., Schoen, V., Specht, K., Grard, B., Blythe, C., Cohen, N., Fox-Kamper, R., Hawes, J., Newell, J., Ponizy, L. (2021). Applying the food-energy-water nexus approach to urban agriculture: From FEW to FEWP (Food-Energy-Water-People). *Urban Forestry and Urban Greening*, 58, 126934. doi: 10.1016/j.ufug.2020.126934.

del Carmen Huesca-Espitia, L., Sánchez-Salas, J.L., Bandala, E.R. (2017). Effect of coat layers in *Bacillus subtilis* spores resistance to photo-catalytic inactivation. *AIMS Geosci.*, 3, 514–525. doi: 10.3934/geosci.2017.4.514.

Castillo-Ledezma, J.H., Salas, J.L.S., López-Malo, A., Bandala, E.R. (2011). Effect of pH, solar irradiation, and semiconductor concentration on the photocatalytic disinfection of Escherichia coli in water using nitrogen-doped TiO2. *Eur. Food Res. Technol.*, 233, 825–834. doi: 10.1007/s00217-011-1579-5.

Castillo-Ledezma, J.H., López-Malo, A., Pelaez, M., Dionysiou, D.D., Bandala, E.R. (2014). Modeling the enhanced photocatalytic solar disinfection of *Escherichia coli* using nitrogen-doped TiO2. *J. Surfaces Interfaces Materials*, 2, 334–342. doi: 10.1166/jsim.2014.1067.

Clurman, A.M., Rodríguez-Narvaez, O.M., Jayarathne, A., De Silva, G., Ranasinghe, M.I., Goonetilleke, A., Bandala, E.R. (2020). Influence of surface hydrophobicity/hydrophilicity of biochar on the removal of emerging contaminants. *Chem. Eng. J.*, 402, 126277. doi: 10.1016/j.cej.2020.126277.

Czepło, F. and Borowski, P.F. (2023). Analysis of legal, political, social and technical aspects concerning enterprises in the energy sector in the era of low-emission economy. *Społeczeństwo i Polityka*, 3(76), 155–176. doi: 10.34765/sp.0323.a09.

Czepło. F. and Borowski, P.F. (2024). Innovation solution in photovoltaic sector. *Energies*, 17(1), 265. doi: 10.3390/en17010265.

Czosnyka, M., Wnukowska, B., Karbowa, K. (2020). Electrical energy consumption and the energy market in Poland during the COVID-19 pandemic. *Prog. Appl. Electr. Eng.*, Koscielisko, 1–5. doi: 10.1109/PAEE50669.2020.9158771.

Dolganova, J., Mikosch, N., Berger, M., Nunez, M., Muller-Frank, A., Finbeiner, M. (2019). The water footprint of European agricultural imports: Hotspots in the context of water scarcity. *Resources*, 8, 141. doi: 10.3390/resources8030141.

Duran-Encalada, J.A., Paucar-Caceres, A., Bandala, E.R., Wright, G.H. (2017). The impact of global climate change on water quantity and quality: A system dynamics approach to the US–Mexican transborder region. *Eur. J. Oper. Res.*, 256, 567–581. doi: 10.1016/j.ejor.2016.06.016.

European Environmental Agency (2018). Ecological status of surface water bodies [Online]. Available at: https://www.eea.europa.eu/themes/water/european-waters/water-quality-and-water-assessment/water-assessments/ecological-status-of-surface-water-bodies [Accessed 1 May 2020].

European Union (2020). Energy policy review. Country Report.

Farias, M.S.D., Schlosser, J.F., Linares, P., Barbieri, J.P., Negri, G.M., Olieira, L.F.V.D., Rudell, I.P.I. (2017). Fuel consumption efficiency of an agricultural tractor equipped with continuously variable transmission. *Cienc. Rural.*, 47. doi: 10.1590/0103-8478cr2016 0814.

Faust, J.E. and Logan, J. (2018). Daily light integral: A research review and high-resolution maps of the United States. *Am. Soc. Hortic. Sci.*, 53, 1250–1257. doi: 10.21273/ hotsci13144-18.

Garcia, T., Boff, L., Regina, C., Barardi, M., Nagl, M. (2019). Inactivation of adenovirus in water by natural and synthetic compounds. *Food Environ. Virol.*, 11, 157–166. doi: 10.1007/s12560-019-09370-8.

Garcia-Gil, A., Martinez, A., Polo-Lopez, M.I., Marugan, J. (2020). Kinetic modeling of the synergistic thermal and spectral actions on the inactivation of viruses in water by sunlight. *Water Res.*, 183, 116074. doi: 10.1016/j.watres.2020.116074.

Goldyn, R., Szpakowska, B., Swierk, D., Domek, P., Buxakowski, J., Doncejewska, R., Baralkiewicz, D., Sajnog, A. (2018). Influence of stormwater runoff on macroinvertebrates in a small urban river and a reservoir. *Sci. Total Environ.*, 625, 743–751. doi: 10.1016/j.scitotenv.2017.12.324.

Gondhalekar, D. and Ramsauer, T. (2017). Nexus city: Operationalizing the urban water-energy-food nexus for climate change adaptation in Munich, Germany. *Urban Clim.*, 19, 28–40. doi: 10.1016/j.uclim.2016.11.004.

Goonetilleke, A., Liu, A., Managi, S., Wilson, C., Gardner, T., Bandala, E.R., Walker, L., Holden, J., Wibowo, M.A., Suripin, S. et al. (2017). Stormwater reuse, a viable option: Fact or fiction? *Econ. Anal. Policy*, 56, 14–17. doi: 10.1016/j.eap.2017.08.001.

Hasterok, D., Castro, R., Landrat, M., Pikori, K., Doepfert, M., Morais, H. (2021). Polish energy transition 2040: Energy mix optimization using grey wolf optimizer. *Energies*, 14, 501. doi: 10.3390/en14020501.

Howden, S.M., Soussana, J.F., Tubiello, F.N., Chetri, N., Dunlop, M., Meinke, H. (2007). Adapting agriculture to climate change. *Proc. Natl. Acad. Sci.*, 104, 19691–19696. doi: 10.1073/pnas.0701890104.

Huesca-Espitia, L.C., Aurioles-López, V., Ramírez, I., Sánchez-Salas, J.L., Bandala, E.R. (2017). Photocatalytic inactivation of highly resistant microorganisms in water: A kinetic approach. *J. Photochem. Photobiol. A Chem.*, 337, 132–139. doi: 10.1016/j.jphotochem. 2017.01.025.

Jager, N.W., Challies, E., Kochskamper, E., Newig, J., Benson, D., Blackstock, K., Collins, K., Ernst, A., Evers, M., Hernandez-Mora, N. (2016). Transforming European water governance? Participation and river basin management under the EU water framework directive in 13 member states. *Water*, 8(4), 156. doi: 10.3390/w8040156.

Jeszke, R. (2019). Klimat dla Polski Polska dla klimatu 1988 – 2018 – 2050 [Online]. Available at: https://cop24.gov.pl/fileadmin/user_upload/files/1._Klimat-dla-Polski-Polska-dla-Klimatu_PL.pdf [Accessed 1 August 2020].

Kaczmarczyk, M., Sowizdzal, A., Tomaszewska, B. (2020). Energetic and environmental aspect of individual heat generation for sustainable development at a local scale–A case study from Poland. *Energies*, 13, 454. doi: 10.3390/en13020454.

Kelemen, A., Munch, W., Poelman, H., Gakova, Z., Dijkstra, L., Torighelli, B. (2009). Regions 2020 the climate change challenge for European regions. Regional Policy, Brussels.

Khan, M.M., Akram, M.T., Janke, R., Qadri, R.W., Al-Sadi, A.M., Farooque, A.A. (2020). Urban horticulture for food secure cities through and beyond COVID-19. *Sustainability*, 12, 9592. doi: 10.3390/su12229592.

Kordana, S. and Slys, D. (2020a). Decision criteria for the development of stormwater management systems in Poland. *Resources*, 9, 20. doi: 10.3390/resources9020020.

Kordana, S. and Slys, D. (2020b). An analysis of important issues impacting the development of stormwater management systems in Poland. *Sci. Total Environ.*, 72, 138711. doi: 10.1016/j.scitotenv.2020.138711.

Korzarcanin, S., Liu, H., Andresen, G.B. (2019). 21st century climate change impacts on key properties of a large-scale renewable-based electricity system, *Joule*, 3, 992–1005. doi: 10.1016/j.joule.2019.02.001.

Kubiak-Wojcicka, K. and Machula, S. (2020). Influence of climate changes in the state of water resources in Poland and their usage. *Geosciences*, 10, 312. doi: 10.3390/geosciences10080312.

Kundzewicz, Z.W. and Matczak, P. (2012). Climate change regional review: Poland. *Wiley Interndisciplinary Rev. Clim. Chang.*, 3, 297–311. doi: 10.1002/wwc.175.

Lewandowska, A., Chodkowska-Miszczuk, J., Rogatka, K. (2020). Development of renewable energy sources in big cities in Poland in the context of urban policy. In *Renewable Energy Sources: Engineering, Technology, Innovation.* Springer, Cham. doi: 10.1007/978-3-030-13888-2_87.

Machac, J., Brabec, J., Volajacek, O. (2020). Development and implementation of the concept of disproportionate cost in water management in central Europe in light of the EU WFD. *Water Alternatives*, 13, 3.

Masson-Delmotte, V., Zhai, P., Pörtner, H.O., Roberts, D., Skea, J., Shukla, P.R., Pirani, A., Moufouma-Okia, W., Péan, C., Pidcock, R. et al. (2018). Global warming of 1.5oC. An IPCC Special Report on the impacts of global warming of 1.5°C above pre-industrial levels and related global greenhouse gas emission pathways, in the context of strengthening the global response to the threat of climate change. Special Report, Intergovernmental Panel on Climate Change.

Montanarella, L. and Panagos, P. (2021). The relevance of sustainable soil management within the European Green Deal. *Land Use Policy*, 100, 104950. doi: 10.1016/j.landusepol.2020.104950.

Newbery, D., Pollitt, M.G., Ritz, R.A. (2018). Market design for a high-renewables European electricity system. *Renew. Sustain. Energy Rev.*, 91, 695–707. doi: 10.1016/j.rser.2018.04.025.

Nistor, M. (2020). Groundwater vulnerability in Europe under climate change. *Quat. Int.*, 547, 185–196. doi: 10.1016/j.quaint.2019.04.012.

Nyga-Lukaszewska, H., Aruga, K., Stala-Szalugaj, K. (2020). Energy security of Poland and coal supply: Price analysis. *Sustainability*, 12, 2541. doi: 10.3390/su12062541.

Piasecki, A. (2020). Water and sewage management issues in rural Poland. *Water*, 11, 625. doi: 10.3390/w11030625.

Piotrowski, P., Baczynski, D., Robal, S., Kopyt, M., Piekarz, M., Polewaczyk, M. (2020). Comprehensive forecast of electromobility mid-term development in Poland and its impacts on power system demand. *Bull. Polish Acad. Sci. Tech. Sci.*, 68, 697–709. doi: 10.24425/bpasts.2020.134180.

Porter, J.R., Xie, L., Challinor, A.J., Cochrane, K., Howden, S.M., Iqbal, M.M., Lobell, D.B., Travasso, M.I. (2014). Food security and food production systems. In *Climate Change 2014: Impacts, Adaptation, and Vulnerability. Part A: Global and Sectoral Aspects. Contribution of Working Group II to the Fifth Assessment Report of the Intergovernmental Panel on Climate Change*. Cambridge University Press, Cambridge, 485–533.

Ptak, E.N., Graversgaard, M., Refsgaard, J.C., Dalgaard, T. (2020). Nitrate management disclosures in Poland and Denmark–Laggards or leaders in water quality protection? *Water*, 12, 2371. doi: 10.3390/w12092371.

Rachmadi, A.T., Kitajima, M., Kato, T., Kato, H., Okabe, S., Sano, D. (2020). Required chlorination doses to fulfill the credit value for disinfection of enteric viruses in water: A critical review. *Environ. Sci. Technol.*, 54, 2068–2077. doi: 10.1021/acs.est.9b01685.

Ramzan, S., Rasool, T., Bhat, R.A., Ahmad, P., Ashraf, I., Rashid, N., ul Shafiq, M., Mir, I.A. (2020). Agricultural soils a trigger to nitrous oxide: A persuasive greenhouse gas and its management. *Environ. Monit. Assess.*, 192, 1–21. doi: 10.1007/s10661-020-08410-2.

Ratajczak, E., Malecka, A., Ciereszko, L., Staszak, A.M. (2019). Mitochondria are important determinants of the aging of seeds. *Int. J. Mol. Sci.*, 20, 1568. doi: 10.3390/ijms20071568.

Rodriguez-Narvaez, O.M., Peralta-Hernandez, J.M., Goonetilleke, A., Bandala, E.R. (2017). Treatment technologies for emerging contaminants in water: A review. *Chem. Eng. J.*, 323, 361–380. doi: 10.1016/j.cej.2017.04.106.

Rodriguez-Narvaez, O.M., Peralta-Hernandez, J.M., Goonetilleke, A., Bandala, E.R. (2019). Biochar-supported nanomaterials for environmental applications. *J. Ind. Eng. Chem.*, 78. doi: 10.1016/j.jiec.2019.06.008.

Royer, T.V. (2020). Stoichiometry of nitrogen, phosphorus, and silica loads in the Mississippi-Atchafalaya River basin reveals spatial and temporal patterns in risk for cyanobacterial blooms. *Linmol. Oceanogr.*, 65, 325–335. doi: 10.1002/lno.11300.

Sanchis, A.G., Pashley, R., Ninham, B. (2019). Virus and bacteria inactivation by CO_2 bubbles in solution. *Clean Water*, 2, 1–9. doi: 10.1038/s41545-018-0027-5.

Satalova, B. and Kenderessy, P. (2017). Assessment of water retention function as tool to improve integrated watershed management (case study of Poprad river basin, Slovakia). *Sci. Total Environ.*, 599–600, 1082–1089. doi: 10.1016/j.scitotenv.2017.04.227.

Schneider, P. and Asch, F. (2019). Rice production and food security in Asian Mega deltas–A review on characteristics, vulnerabilities and agricultural adaptation options to cope with climate change. *J. Agron. Crop Sci.*, 206, 491–503. doi: 10.1111/jac.12415.

Seinn, M.U. and Ahmand, M.M. (2015). Farmers' adaptation to rainfall variability and salinity through agronomic practices in lower ayeyarwady delta, Myanmar. *J. Earth Sci. Clim. Chang.*, 6, 258. doi: 10.4172/2157-7617.1000258.

Soika, M., Kozlowski, M., Kesicka, B., Wrozynski, R., Stasik, R., Napierala, M., Liberacki, D. (2020). The effect of climate change on controlled drainage effectiveness in the context of groundwater dynamics, surface, and drainage outflows. Central-Western Poland case study. *Agronomy*, 20, 625. doi: 10.3390/agronomy10050625.

Stein, E.W. (2021). The transformative environmental effects large-scale indoor farming may have on air, water, and soil. *Air, Soil Water Research*, 14, 117862212199581.

Szabo, J. and Fabok, M. (2020). Infrastructures and state-building: Comparing the energy politics of the European Commission with the governments of Hungary and Poland. *Energy Policy*, 138, 111253. doi: 10.1016/j.enpol.2020.111253.

Topal, D., Hatvani, I.G., Kern, Z. (2020). Refining projected multidecadal hydroclimate uncertainty in East-Central Europe using CMIP5 and single-model large ensemble simulations. *Theor. Appl. Climatol.* doi: 10.1007/s00704-020-03361-7.

U.S. Department of State (2020). Investment climate statements: Poland [Online]. Available at: https://www.state.gov/reports/2020-investment-climate-statements/poland/.

Van Vuuren, D.P., Bijl, D.L., Bogaart, P., Stehfest, E., Biemans, H., Dekker, S.C., Doelman, J.C. Gernaat, D.E.H.J., Harmsen, M. (2019). Integrated scenarios to support analysis of the food-energy-water nexus. *Nat. Sustain.*, 2, 1132–1141. doi: 10.1038/s41893-019-0418-8.

Vogel, E. and Meyer, R. (2018). Climate change, climate extremes, and global food production – Adaptation in the agricultural sector. In *Resilience : The Science of Adaption to Climate Change*, Zommers, Z. and Alverson, K. (eds). Elsevier, Amsterdam, 31–49. doi: 10.1016/B978-0-12-811891-7.00003-7.

Wang, C., Horby, P.W., Hayden, F.G., Gao, G.F. (2020). A novel coronavirus outbreak of global health concern. *Lancet*, 395, 470–473.

Wibig, J. and Jakusik, E. (2012). *Warunki klimatyczne i oceanograficzne w Polsce i na Bałtyku Południowym: Spodziewane zmiany i wytyczne do opracowania strategii adaptacyjnych w gospodarce krajowej*. IMGW-PIB, Warsaw.

World Energy Council (2016). Solar. In *World Energy Resources | 2016.* World Energy Council, London [Online]. Available at: https://www.worldenergy.org/assets/images/imported/2016/10/World-Energy-Resources-Full-report-2016.10.03.pdf.

WWP (2017). Wastewater: The untapped resource. United Nations World Water Development Report, Paris.

Zawalinska, K., Kinnunen, J., Gradziuk, P., Celinska-Janowicz, D. (2020). To whom should we grant a power plant? *Economic Effects of Investment in Nuclear Energy in Poland, Energies*, 13, 2687. doi: 10.3390/en13112687.

Zhang, C., Xu, L., Xu, P., Wang, X.C. (2018). Elimination of viruses from domestic wastewater: Requirements and technologies. *World J. Microbiol. Biotechnol.*, 32, 69. doi: 10.1007/s11274-016-2018-3.

Zhang, X., Lardizabal, A., Silverman, A.I., Vione, D., Kohn, T., Nguyen, T.H., Guest, J.S. (2020). Global sensitivity analysis of environmental, water quality, photoreactivity, and engineering design parameters in sunlight inactivation of viruses. *Environ. Sci. Technol.*, 54, 8401–8410. doi: 10.1021/acs.est.0c01214.

3

Management and Analysis of the Sustainable Initiatives in the Agri-Food Supply Chain in India

Owing to the poor supply chain process, nearly 40% of the food products produced are being wasted or spoiled in India. The wastage is happening at every stage of the supply chain, which starts from harvesting, transporting, processing, packaging and consuming. As a result, this is adding a cost of around one lakh crore rupees to the country's expenditure every year. Moreover, the growing population demands more food and other agriculture-based products. Therefore, there is a major concern to focus on the quality, safety and hygiene of food-based supply products. In view of the above-mentioned aspects, this is the need of the hour, which will provide efficient agri-food supply chain (AFSC) practices that help in reducing the wastage of food products at various stages of the supply chain process. This study has identified various sustainable initiatives through pertinent literature review and domain specialists' inputs, which will emphasize the proper usage of resources to optimize the environmental and socio-economic factors, which are creating a major impact on the agricultural food product business. These initiatives are analyzed using the fuzzy analytic hierarchy process (FAHP) approach. The study findings will provide managerial and policy implications to key stakeholders that will enable the suppliers and consumers in the areas of effective utilization of the available resources and to reduce the wastage of food products.

Chapter written by Chandra Prakash GARG, Munnaluru Sai Venkata YESWANTH and Archana SHARMA.

3.1. Introduction

Agriculture is observed as the primary source of income for Indian people in many rural parts of the country. The vegetables and other spices that are being cultivated in these regions are being supplied to consumers in different parts of the nation. In earlier times, people used to consume the food products that are available only in that area. As the food-consuming habits of individuals gradually changed, this caused them to look after a variety of food products produced in various parts of the country. However, the supply chain of perishable vegetables, other agricultural food products and cultivated food products have huge wastages or become foul, despite reaching the consumers in the same state in which the product is produced.

According to the economic survey 2017–2018, 50% of the total workforce in the country is in the agriculture sector. Also, the sector is contributing 17–18% to the nation's GDP and it is estimated that the workforce may fall to 25.7% by 2050. This is because of the non-profitable scenario of the agri-food supply chain industry at many stages before the product reaches the final consumer. According to recent statistics, in Indian demography, 70% of the rural population is having agriculture as their primary source of livelihood. The country is recognized as one of the largest producers of food products, standing second in the world. It is observed from the statistics made by the UN that approximately 40% of the food products produced are being wasted or spoiled due to lack of effective supply chain management practices. This wastage is happening at every stage of the supply chain process. However, more than 65% of food wastage could be avoided and safely redirected to human consumption (Liu et al. 2022).

Considering the present challenges in the area of AFSCM (Agri-Food Supply Chain Management), this work has identified various sustainable initiatives for bringing sustainability to the supply chain. Then, these initiatives are evaluated with the fuzzy analytic hierarchy process (FAHP) method. This fuzzy-based multi-criteria decision-making framework resolves the uncertainty and impreciseness in the decision-making process.

The latter part of this chapter will deal in a prescribed manner. Section 3.2 provides an incisive literature review about AFSC and will provide some scope for identifying the critical factors. The objective of this study, along with research issue definition, is given in section 3.3. The methodology of the study, i.e. the FAHP procedure, is discussed in section 3.4. The ranking of identified critical factors of the study is presented in section 3.5. The conclusion is discussed in the final section 3.6.

3.2. Literature review

3.2.1. *Sustainable initiatives in the agri-food supply chain (AFSC)*

This supply chain consists of many activities like acquiring raw materials, manufacturing or processing of food products, post harvesting, storage of products according to their nature, stacking and distribution and making the product available at markets (Gokarn and Kuthambalayan 2017). The main supplier is the farmer who is producing the food product and the main consumer is the end purchaser. At the Indian market, the AFSC system is determined as a cluster of issues such as dominance of small and marginal farmers, fragmented supply chains, the absence of small-scale economies, low level of value addition and inadequacy of marketing infrastructure, etc. (Somashekhar et al. 2014). The major issue in this supply chain is to maintain the coordination among all the stakeholders involved in the process (Garg et al. 2022b). The major constraints are poor transportation facilities, unavailability of large cold storage facilities and no clear policy guidelines from government and other institutions involved. The AFSC has long lead times and is not properly aligned. There is huge uncertainty between the demand and supply of the products (Laosirihongthonga et al. 2013). The food supply chain in India is witnessing massive wastage and inefficiencies contributing to 30% of India's fruit and vegetables being wasted. This improper supply chain is resulting in the periodic shortage of food items that are being used by the consumer as a part of their diet (Somashekhar et al. 2014).

The AFSC in India is highly segmented. The more food produced, the more the intermediaries along the SCM (Supply Chain Management), because of their vital role. Major incarceration in Indian AFSCM means very meagre transport facilities, not having cold storages on a larger scale, not having proper guidelines from the government on the supply chain, large fragmented intermediaries and small farmers who do not have much knowledge on this supply chain. Existing supply chains are very much extensive and are influenced by many intermediaries such as facilitators, distributors, wholesalers, retailers, commission agents and local marketers. Farmers in India receives one-third to one-half of the final price in the case of fruits and vegetables. Many perishable products such as milk, green vegetables and fruits are being wasted in India without reaching consumers because of the lack of infrastructure and mishandling of the products at these stages. This mishandling is the serious issue to be addressed by all of the authorities involved in supply chain management. The sustainability framework of AFSCM has emanated as a global problem and needs an effective solution to the rise in food demand (Barbosa 2021). Although agriculture is a primary source of employment in rural India, it suffers from inefficiencies in SCM (Kumar and Choubey 2021). Scarcity in resources that are naturally available will lead to the declining rate of supply of the food products, which will be a serious issue in the coming future. Wastage of food, losses due to ineffective supply chains, lack of infrastructure facilities and poor capacity planning are the key indicators driving the sustainability into a negative slope (Kayıkcı et al. 2021). Sustainability initiatives are defined for the purpose of maintaining sustainable harmony along the SCM of agricultural food. This literature review boosted the study for identifying the factors or key enablers for creating a perfect supply chain in the Indian food SCM process. Proper usage of resources available, utilization of storage houses and the efficient transportation of food products would create a better chance to conceive the ecological, connected issue which will be disturbing the stable development in India. For the fruition of social needs, ecological impacts, financial advantages, long-term stability and availing new business chances, AFSC in India has to imply some of the factors given by the European Union (EU) (Humphrey and Memedovic 2006).

3.3. Problem definition

S. no.	SI	Code	References
1	Subsidies and support by different agencies for SI	SI1	Beske et al. (2014) and Luthra et al. (2018)
2	Regulation of different governmental institutes and government authorities	SI2	Ding et al. (2019)
3	Assimilating the significance of final end customer and stakeholders prospects	SI3	Grimm et al. (2014) and Gokarn et al. (2017)
4	Proper allocation of resources and sharing of information along the supply chain	SI4	Bitzer et al. (2008) and Kumar and Choubey (2021)
5	Coordination in planning and capacity building in delivering the products	SI5	Grimm et al. (2014) and Kayıkcı et al. (2021)
6	Auditing and controlling of ongoing supply chain activities	SI6	Luthra et al. (2017) and Barbosa (2021)
7	Organizational competitive advantage	SI7	Sengar et al. (2018) and Liu et al. (2022)

Table 3.1. *Identified sustainable initiatives (SI) of AFSCM*

There are huge wastages and losses of agri-food items in the various stages of the supply chain process in India. There is a need to recognize and assess the critical sustainable initiatives which can support reducing wastage. The key initiatives in the study have identified to elevate the sustainable procedures in AFSCM. The research would help in reducing the wastage of food products to a short time period. These initiatives have been identified with the consensus of supply chain industry experts of around 25 years of experience. Also, this study would evaluate and prioritize the identified initiatives for the purpose of creating a sustainable environment at the AFSC. Table 3.1 shows identified critical factors which will enact as the enablers for the framework of sustainability.

3.4. Solution methodology

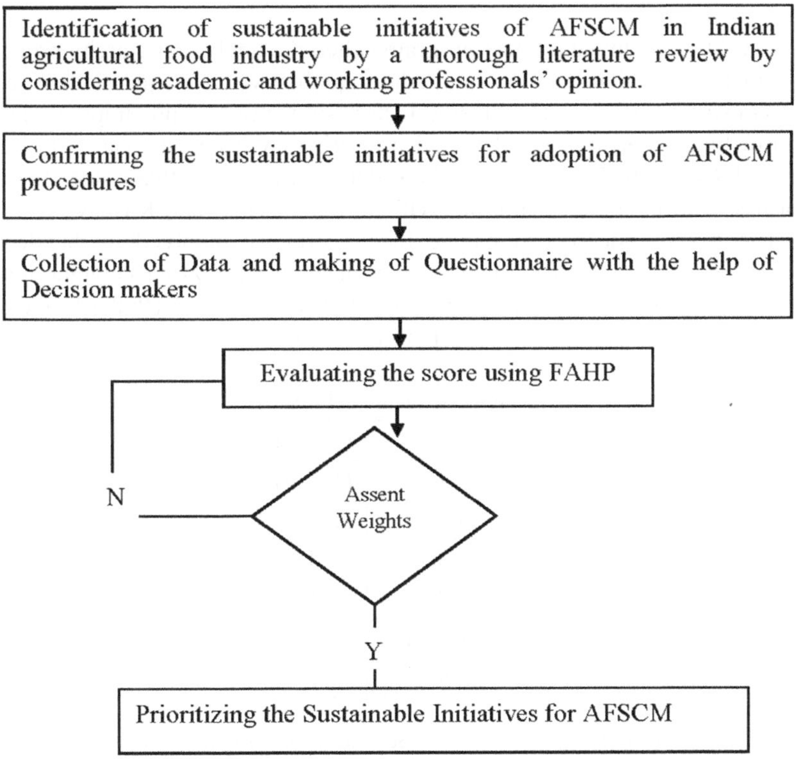

Figure 3.1. *Defined FAHP methodology of SI assessment of AFSCM*

In this study, seven SI of AFSCM were identified to develop structured initiatives (see Table 3.1). This study used the FAHP to assign suitable ranking and assess SI for successful implementation for further improvements in the process of attaining sustainability in AFSCM. The analytic hierarchy process (AHP) is a quantitative technique that deals with complex problems. This technique has some unique features such as offering better results with regards to decisions related to pairwise comparison of the vectors, uses less information and involves decision-makers as compared to other techniques such as VIKOR, TOPSIS and DEMATEL. This methodology was carried in the form of sequential steps, as discussed below, to obtain the hierarchy of SI. This is shown schematically in the form of a flow chart in Figure 3.1.

3.4.1. *Fuzzy AHP method*

Satty (1980) developed a linear program called the analytic hierarchy process for decision-making, which can accommodate qualitative and quantitative criteria simultaneously. However, this method has subjectivity and impreciseness in decision-making when it would be used in a crisp environment (Garg et al. 2017; Kumar and Garg 2017). Therefore, in order to overcome this issue, the fuzzy concept has been integrated with the AHP approach (Garg and Kashav 2019; Garg 2020). The fuzzy concept can reduce the uncertainty involved in the decision-making process (Garg 2016). Many studies have used fuzzy triangular numbers (TFNs) along with the AHP approach, as listed in Table 3.3 (Prakash and Barua 2015, 2016a, 2016b, 2016c, 2016d; Kumari et al. 2018).

Linguistic term	TFN
Equally	(1, 1, 1)
Very low	(1, 2, 3)
Low	(2, 3, 4)
Average	(3, 4, 5)
High	(4, 5, 6)
Very high	(5, 6, 7)
Extreme	(7, 8, 9)

Table 3.2. *TFN representation. Source: Mahtani and Garg (2018)*

This study applied Chang's extent analysis as follows. Each criterion score c_i is obtained using TFNs.

Step 1: The fuzzy synthetic extent value is obtained for each criterion

$$E_i = \sum_{j=1}^{m} S_{c_i}^{j} \times \left[\sum_{i=1}^{n}\sum_{j=1}^{m} S_{c_i}^{j}\right]^{-1} i = 1, 2, 3, 4, 5 \ldots\ldots. n \text{ and } j = 1, 2, 3, 4, 5 \ldots\ldots.. m \quad [3.1]$$

Step 2: The possibility degree of

$E_2 = (x_2, y_2, z_2) \geq E_1 = (x_1, y_1, z_1)$ is obtained as

$$V (E_2 \geq E_1) = \sup_{b \geq a}[\min(\mu_{E_1}(a), \mu_{E_2}(b)] \quad [3.2]$$

where a, b represents the membership value of the fuzzy set.

Step 3: The minimum possibility degree is determined by comparing 1 convex fuzzy sets with E_i (i= 1,2,…..,l) are computed using

$$V (E \geq E_1, E_2, \ldots\ldots., E_l) \qquad i = 1,2,\ldots..,l$$

$$= V [(E \geq E_1) \text{ and } (E \geq E_2) \text{ and } \ldots\ldots.. \text{and } (E \geq E_l)]$$

$$V = \min V (E \geq E_i) \qquad i = 1,2,\ldots..,l \quad [3.3]$$

Assuming $d'(S_i) = \min V (E_i \geq E_l)$

For $l = 1, 2, \ldots n, l \neq i$, hereafter criteria weights are calculated in equation [3.3] as

$$W' = (d'(S_1), d'(S_2), \ldots\ldots., d'(S_m))^T$$

Step 4: Then, normalized weights of the criteria are obtained, as described in equation [3.4]

$$W = (d(S_1), d(S_2), \ldots\ldots., d(S_m))^T \quad [3.4]$$

The domain group is allocated the linguistic rating to assess the SI of AFSCM using Table 3.2. The final rating is highlighted in Table 3.3. Then, FAHP analysis is applied, as discussed above, in the research outline section and the final ranking of the SI is highlighted in Table 3.4.

	SI1	SI2	SI3	SI4	SI5	SI6	SI7
SI1	(1, 1, 1)	(0.25, 0.33, 0.5)	(3, 4, 5)	(1, 2, 3)	(0.2, 0.25, 0.33)	(0.25, 0.33, 0.5)	(3, 4, 5)
SI2	(2, 3, 4)	(1, 1, 1)	(0.33, 0.5, 1)	(2, 3, 4)	(2, 3, 4)	(3, 4, 5)	(0.33, 0.5, 1)
SI3	(0.2, 0.25, 0.33)	(1, 2, 3)	(1, 1, 1)	(1, 2, 3)	(3, 4, 5)	(0.33, 0.5, 1)	(0.25, 0.33, 0.5)
SI4	(0.33, 0.5, 1)	(0.25, 0.33, 0.5)	(0.33, 0.5, 1)	(1, 1, 1)	(0.25, 0.33, 0.5)	(3, 4, 5)	(2, 3, 4)
SI5	(3, 4, 5)	(0.25, 0.33, 0.5)	(0.2, 0.25, 0.33)	(2, 3, 4)	(1, 1, 1)	(1, 2, 3)	(0.33, 0.5, 1)
SI6	(2, 3, 4)	(0.2, 0.25, 0.33)	(1, 2, 3)	(0.2, 0.25, 0.33)	(0.33, 0.5, 1)	(1, 1, 1)	(0.25, 0.33, 0.5)
SI7	(0.2, 0.25, 0.33)	(1, 2, 3)	(2, 3, 4)	(0.25, 0.33, 0.5)	(1, 2, 3)	(2, 3, 4)	(1, 1, 1)

Table 3.3. *Matrix for SI in TFNs. Source: FAHP analysis*

S. no.	SI	Code	Final weights	Rank
1	Subsidies and support by different agencies for SI	SI1	0.1567	2
2	Regulation of different governmental institutes and government authorities	SI2	0.1919	1
3	Assimilating the significance of the final end customer and stakeholders prospects	SI3	0.1347	5
4	Proper allocation of resources and sharing of information along the supply chain	SI4	0.1269	6
5	Coordination in planning and capacity building in delivering the products	SI5	0.1475	4
6	Auditing and controlling of the ongoing supply chain activities	SI6	0.0873	7
7	Organizational competitive advantage	SI7	0.1551	3

Table 3.4. *Final ranking of SI in AFSCM. Source: FAHP analysis*

3.5. Results and discussion

The ranking of SI is assigned by considering the final weights. After the application of the FAHP, the sequence of SI is as follows: SI2 > SI1 > SI7 > SI5 >

SI3 > SI4 > SI6; according to the weights, they are arranged in descending order (see Table 3.4). The results of the FAHP analysis show that SI2, i.e. regulation of different governmental institutes and government authorities, obtained the first rank with a weightage value of 0.1919. This indicates that Indian agencies need to interfere in the AFSC process and operations, since the regulation is the most important driver to implement various policy initiatives. Similarly, governments need to bring up directives and rules post-harvesting pertaining to agro-food-based products. The supply chain partners or middlemen are responsible for value creation from agro-food-based products post-harvesting. Owing to money-making goals, these middlemen keep huge stocks of these agro-food-based products, leading to unnecessary price hikes of these items and wastages. If there was a strict government policy, it would limit the problem of overstocking and further reduce food wastage. The second next enabler for the sustainable improvement of AFSCM is SI1 (0.1567), subsidies and support by different agencies for SI. It suggests that the government must provide subsidies and support for the development of warehouses and infrastructure for food stocks. Financial agencies and institutions provide cheaper loans for the development of food item-related warehouses. The central and state governments would provide necessary support and schemes for cold chain logistics development in various areas. The third enabler is SI7 (0.1551), which is a healthy organizational competitive advantage. In India, the aggregator, called middlemen, usually purchase these freshly produced products from farmers. There is no such formal competition among these middlemen, which hampers the bargaining power of the farmers. In such a situation, farmers keep the stock of freshly produced items in a traditional manner without properly maintaining specific atmospheric and temperature control conditions, which deteriorates these perishable items and leads to huge wastages. Therefore, this initiative needs specific attention from the key stakeholders. Also, the number of intermediaries should be constrained to create a profitable corridor to the farmers who are involved in the supply chain. Next, the weighted enabler in the hierarchy is SI5 (0.1475); a perfect coordination in planning and capacity building in delivering the products. There should be mutual coordination in planning the movement of food products in terms of capacity building, not exceeding the supply of food products than demand at the time of delivery to the customers, which can be resolved by intervening with the help of technology aspects, both at planning and capacity building stages. The sustainable driver such as understanding the significance of customer and stakeholders requirements SI3 (0.1347) stood fifth in the rankings. The norms and policies of customer demand and supply chain partners are clearly confined, thereby knowing the interest of stakeholders' requirements in the supply chain for sustainable agricultural products. By knowing the interests in a significant manner, the required resources are to be allocated according to their policies. This can be done by sharing every pinch of information among the players involved in the supply chain. SI4 (0.1269), proper allocation of required materials and sharing of details along the supply chain, is the enabler at the sixth position. The final enabler SI6 (0.0873) is auditing and controlling ongoing supply chain activities, which stands last in the hierarchy of SI. The final required sustainable enabler or success factor in any industry is through auditing; this activity has to be done on a defined time period basis by the regulating bodies, to look after and control any violations occurring in the ongoing supply chain of agriculture food products.

These initiatives need to be implemented on a priority basis for better-driven agricultural food product supply chain management at the market.

3.6. Sensitivity analysis

The sensitivity analysis was performed to showcase the validity of the proposed method. Sensitivity experiments were performed by changing the highest weight initiative, i.e. SI2 from 0.1 to 0.5; then, the respective weights of the other SI were obtained, as highlighted in Table 3.5 (Rawat and Garg 2023; Garg et al. 2022a, 2022b). To show the effect on the ranking of the SI, an incremental change in the weightage value of SI2 was made from 0.1 to 0.5. Thereafter, the final ranking of the SI was obtained, as highlighted in Table 3.6. It shows that when the SI2 weightage value is 0.1, the first rank is obtained by SI1 and the last rank is obtained by SI2. Hereafter, when the SI2 weightage value varies from 0.2 to 0.5, then the first rank is obtained by SI2 and the last rank is obtained by SI6 in all experiments. The results of the sensitivity analysis, as highlighted in Figure 3.2, show that the model used has less variation. Hence, the model is robust enough to evaluate the SI of AFSCM.

SI	Weights of the SI					
SI1	0.1720	0.1567	0.1554	0.1387	0.1220	0.1054
SI2	0.1000	0.1919	0.2000	0.3000	0.4000	0.5000
SI3	0.1500	0.1347	0.1334	0.1167	0.1000	0.0834
SI4	0.1422	0.1269	0.1256	0.1089	0.0922	0.0756
SI5	0.1628	0.1475	0.1462	0.1295	0.1128	0.0962
SI6	0.1025	0.0872	0.0859	0.0692	0.0525	0.0359
SI7	0.1704	0.1551	0.1538	0.1371	0.1204	0.1038
Total	1	1	1	1	1	1

Table 3.5. *SI scores when increasing the weightage value of SI2*

SI	SI ranking when varying the weight of SI2					
	0.1	**Normalized (0.1919)**	**0.2**	**0.3**	**0.4**	**0.5**
SI1	1	2	2	2	2	2
SI2	7	1	1	1	1	1
SI3	4	5	5	5	5	5
SI4	5	6	6	6	6	6
SI5	3	4	4	4	4	4
SI6	6	7	7	7	7	7
SI7	2	3	3	3	3	3

Table 3.6. *SI ranking when the SI2 score varies*

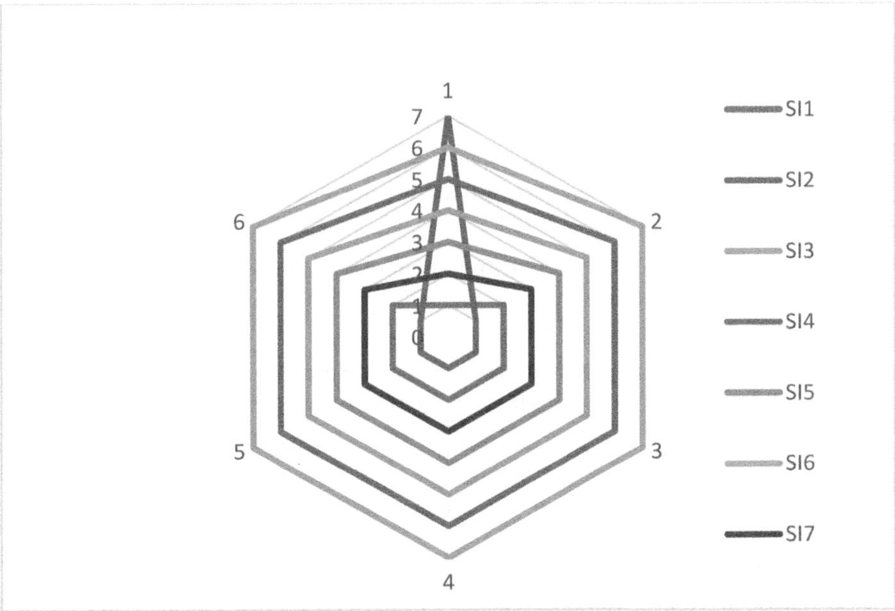

Figure 3.2. *Results of the sensitivity analysis. For a color version of this figure, see www.iste.co.uk/koubaa/optimization.zip*

3.7. Conclusion

This study would raise awareness about the hierarchy of sustainable enablers or initiatives in the area of agricultural food supply chain management. As the study was made considering the exclusive literature review, more scope for the fieldwork will help researchers draw many more key enablers in the supply chain industry. Also, these key enablers will improve the situation of supply chain management in an Indian context. These enablers will certainly be helpful to reduce the wastage to a greater extent. This improvement in the reduction of wastage will add to the economic status of India's GDP; the tool used in the study is confined to determine the rankings, thereby giving priority based on the weightages. This study can be extended by identifying more initiatives considering the micro-supply chain management systems, and evaluate them with the help of other research techniques such as MAUT, TOPSIS and DEMATEL.

3.8. References

Barbosa, M.W. (2021). Uncovering research streams on agri-food supply chain management: A bibliometric study. *Global Food Security*, 28, 100517.

Beske, P., Land, A., Seuring, S. (2014). Sustainable supply chain management practices and dynamic capabilities in the food industry: A critical analysis of the literature. *International Journal of Production Economics*, 152, 131–143.

Bitzer, V., Francken, M., Glasbergen, P. (2008). Intersectoral partnerships for a sustainable coffee chain: Really addressing sustainability or just picking (coffee) cherries? *Global Environmental Change*, 18(2), 271–284.

Ding, H., Fu, Y., Zheng, L., Yan, Z. (2019). Determinants of the competitive advantage of dairy supply chains: Evidence from the Chinese dairy industry. *International Journal of Production Economics*, 9, 360–373.

Garg, C.P. (2016). A robust hybrid decision model for evaluation and selection of the strategic alliance partner in the airline industry. *Journal of Air Transport Management*, 52, 55–66.

Garg, C.P. (2020). Evaluation of the airport service quality indicators using fuzzy AHP framework: An Indian perspective. *International Journal of Services and Operations Management*, 37(3), 320–342.

Garg, C.P. and Kashav, V. (2019). Evaluating value creating factors in greening the transportation of Global Maritime Supply Chains (GMSCs) of containerized freight. *Transportation Research Part D: Transport and Environment*, 73, 162–186.

Garg, C.P., Sharma, A., Goyal, G. (2017). A hybrid decision model to evaluate critical factors for successful adoption of GSCM practices under fuzzy environment. *Uncertain Supply Chain Management*, 5(1), 59–70.

Garg, C.P., Görçün, Ö.F., Kundu, P., Küçükönder, H. (2022a). An integrated fuzzy MCDM approach based on Bonferroni functions for selection and evaluation of industrial robots for the automobile manufacturing industry. *Expert Systems with Applications*, 213(A), 118863.

Garg, C.P., Kashav, V., Wang, X. (2022b). Evaluating sustainability factors of green ports in China under fuzzy environment. *Environment, Development and Sustainability*, 25(8), 7795–7821.

Gokarn, S. and Kuthambalayan, T.S. (2017). Analysis of challenges inhibiting the reduction of waste in food supply chain. *Journal of Cleaner Production*, 168, 595–604.

Grimm, J.H., Hofstetter, J.S., Sarkis, J. (2014). Critical factors for sub-supplier management: A sustainable food supply chains perspective. *International Journal of Production Economics*, 152, 159–173.

Humphrey, J. and Memedovic, O. (2006). Global value chains in the agrifood sector. Working paper, UNIDO.

Kayıkcı, Y., Gözaçan, N., Lafcı, Ç., Kazançoğlu, Y. (2021). A conceptual framework for food loss and waste in agri-food supply chains: Circular economy perspective. In *Challenges and Opportunities of Circular Economy in Agri-Food Sector*, Mor, R.S., Panghal, A., Kumar, V. (eds). Springer, Singapore, 41–53.

Kumar, M. and Choubey, V.K. (2021). Modeling the causes of post-harvest loss in the agri-food supply chain to achieve sustainable development goals: An ISM approach. In *Challenges and Opportunities of Circular Economy in Agri-Food Sector*, Mor, R.S., Panghal, A., Kumar, V. (eds). Springer, Singapore, 133–149.

Kumar D. and Garg C.P. (2017). Evaluating sustainable supply chain indicators using Fuzzy AHP: Case of Indian automotive industry. *Benchmarking: An International Journal*, 24(6), 1742–1766.

Kumar, D. and Rahman, Z. (2013). Modelling relationship marketing strategies for sustainability adoption. *International Journal of Intercultural Information Management*, 3(4), 277–291.

Kumari, A., Garg, C.P., Singh, S. (2018). Modelling the critical success factors of women entrepreneurship using Fuzzy AHP framework. *Journal of Entrepreneurship in Emerging Economies*, 10(1), 81–116.

Laosirihongthonga, T., Punnakitikashemb, P., Adebanjo, D. (2013), Improving supply chain operations by adopting RFID technology: Evaluation and comparison of enabling factors. *Production Planning & Control: The Management of Operations*, 24(1), 90–109.

Liu, C., Jiang, H., Badulescu, D., Bac, D.P. (2022). Achieving zero hunger goal through minimizing waste in food supply chain: Evidence from Asian emerging region. *Sustainability*, 14(10), 5930.

Luthra, S., Govindan, K., Kannan, D., Mangla, S.K., Garg, C.P. (2017). An integrated framework for sustainable supplier selection and evaluation in supply chains. *Journal of Cleaner Production*, 140, 1686–1698.

Luthra, S., Mangla, S.K., Shanker, R., Garg, C.P., Jakhar, S.K. (2018). Modelling critical success factors for sustainability initiatives in supply chains in Indian context using Grey-DEMATEL. *Production Planning & Control*, 29(9), 705–728.

Mahtani, U.S. and Garg, C.P. (2018). An analysis of key factors of financial distress in airline companies in India using fuzzy AHP framework. *Transportation Research Part A: Policy and Practice*, 117, 87–102.

Prakash, C. and Barua, M.K. (2015). Integration of AHP-TOPSIS method for prioritizing the solutions of reverse logistics adoption to overcome its barriers under fuzzy environment. *Journal of Manufacturing System*, 37, 599–615.

Prakash, C. and Barua, M.K. (2016a). An analysis of integrated robust hybrid model for third-party reverse logistics partner selection under fuzzy environment. *Resources, Conservation & Recycling*, 108, 63–81.

Prakash, C. and Barua, M.K. (2016b). A combined MCDM approach for evaluation and selection of third-party reverse logistics partner for Indian electronics industry. *Sustainable Production and Consumption*, 7, 66–78.

Prakash, C. and Barua, M.K. (2016c). A robust multi-criteria decision making framework for evaluation of the airport service quality enablers for ranking the airports. *Journal of Quality Assurance in Hospitality & Tourism*, 17(3), 351–370.

Prakash, C. and Barua, M.K. (2016d). A multi criteria decision making approach for prioritizing reverse logistics adoption barriers under fuzzy environment: Case of Indian electronics industry. *Global Business Review*, 17(5), 1–18.

Prakash, C. and Barua, M.K. (2017). Flexible modelling approach for evaluating reverse logistics adoption barriers using fuzzy AHP and IRP framework. *International Journal of Operational Research*, 30(2), 151–171.

Rawat, A. and Garg, C.P. (2023). Modeling the strategies to accelerate the natural gas business market growth in a developing country. *Journal of Business & Industrial Marketing*, 38(5), 1116–1134.

Saaty, T. (1980). The analytic hierarchy process (AHP) for decision-making. In *Kobe, Japan*, 1–69.

Sengar, V.S., Garg, C.P., Raju, T.B. (2018). Assessment of sustainable initiatives in Indian ports using AHP framework. *International Journal of Business Excellence*, 16(1), 110–126.

Somashekhar, I.C., Raju, J.K., Patil, H. (2014). Agriculture supply chain management: A scenario in India. *TIJRP RJSSM Research Journal of Social Science and Management*, 4(7), 89–99.

Vishwakarma, V., Prakash, C., Barua, M.K. (2016). A fuzzy-based multi criteria decision making approach for supply chain risk assessment in Indian pharmaceutical industry. *International Journal of Logistics Systems and Management*, 25(2), 245–265.

Vishwakarma, V., Prakash, C, Barua, M.K. (2019). Modelling the barriers of Indian pharmaceutical supply chain using fuzzy AHP. *International Journal of Operational Research* 34(2), 240.

4

Inventory Management Activities Determinants Post IT Applications in Indian Sugar Supply Chain

The issues for managing inventory activities using secondary sources and expert panel members are identified. Respondents developed a suitable relationship between the identified issues to facilitate IT applications for managing inventory. A five-point Likert scale was used to check the relevance level, and a total of 340 respondents were from mill managers, distributors, IT companies and large-sized sugar-consuming companies (LSSCC). A field survey was conducted for the identified issues in both structured and unstructured ways to check the suitability of IT applications related to inventory management activities. The results are statistically significant and correlate with each other, which helps facilitate IT applications. Therefore, some recommended IT solutions with certain implications are provided for different inventory management activities.

4.1. Introduction

The IT industry (Gupta and Basole 2020) is a rapidly emerging industry in India, which developed from 150 million US dollars in 1990–1991 to 87 billion US dollars in 2008–2009 (Desai 2008; Chamarbagwala and Sharma 2010). Several multinational IT companies took initiatives in almost all types of segments for providing solutions for day-to-day functioning and applying them. With the development of the IT industry in India, IT-empowered administrations have moreover extended. These managements are presented by different IT applications

---

Chapter written by Rupesh KUMAR, Deepak BANGWAL, Pradeep CHAUHAN and Githa S. HEGGDE.

to advance easiness and correctness. A portion of the IT gadgets are used to encourage viable agrarian arranging and advancement, for example, data warehousing (databases), networking (Internet, intranet and extranet), GIS (Geographical Information Systems), remote sensing data application, e-commerce and e-governance (Moni 2003; Fatemeh et al. 2020).

The use of e-applications in the business-to-business (B2B) and business-to-customers (B2C) sectors is quickly emerging among many business associations (Kalakota and Robinson 1999; Goodridge 2000; Harwood 2001). Information and communication technology (ICT) has driven many developments and establishment of agriculture areas in India. Many Internet portals such as ITC e-Choupal, Agmarket, Gyandoot, Warana Wired Village Project and Indiagriline are involved in e-commerce practices in different regions of the nation (Goswami et al. 2005; Kansara et al. 2022). Horticulture has some extraordinary imperatives identified with the adaption of IT.

Synchronized inventory chains can offer an upper hand in the current globalizing business economy. ICT and the use of e-applications have changed the impression of production network rehearses. By using e-applications in the B2B and B2C markets, the inventory network has become client-driven as opposed to item-arranged (Harwood 2001; Bangwal et al. 2022). In B2B markets, e-applications help in giving information everywhere, and permit organizations to synchronize their exercises such as bundling, procurement, transportation, creation arranging and advertising (Timmers 1999). In B2C markets, the stream of information in the store network using the web helps in the increment of production network perceivability (Goodridge 2000). The store network executives are a significant viewpoint for the sustenance of the whole sugar industry. By using IT applications, various types of board exercises such as appropriation, stock, transportation coordination and vendors can be monitored effectively and successfully. Numerous issues developed and identified with stock administration and the potential effect of an appropriate IT application/arrangement in encouraging these issues were reviewed.

4.1.1. *Objective of the study*

The following are the precise objectives of the study:

– to identify issues related to inventory management activities of sugar industries;

– to explore the impact of IT applications that can facilitate inventory management.

The study was conducted in two parts. The first part identifies the relevant issues belonging to manage inventory activities. Therefore, the survey is conducted with the respondents involving millers, distributors, IT companies and large-sized sugar-consuming companies (LSSCC) and through secondary sources. The second part consisted of the application of factor analysis (Clark and Bowles 2018) to arrive at the issues that govern inventory management activities. A regression method was used to establish a relationship between such issues and facilitation by IT applications, and then a hypothesis was tested using ANOVA to check the significant differences among the identified issues.

4.2. Literature review

The literature reveals the frequent use of e-applications for stock administration in the parts such as car, fabricating, horticulture, etc. At the same time, few scholastic studies were identified, which dealt with the administration of stock exercises of the Indian sugar inventory network.

Ross (1997), Palevich (1999) and Simatupang and Sridharan (2002) stated that the production network of the board is the time taken to move and put away the crude materials, work-in-process stock and completed merchandise from one point to the end-shopper. Keith Oliver (Booz Allen consultant) characterized supply chain management (SCM) as a procedure of actualizing, arranging and controlling the tasks to satisfy the necessities or requirements of a client in a proficient manner. Heckmann et al. (2003), Jacoby (2009) and Blanchard (2010) clarified that the term supply chain management was authored first by the US Industry Consultant in the mid-1980s.

Crouch (2003) set forward an assortment of some investigations with respect to the execution of online business in a store network. He reasoned that associations could profit as cost reserve funds and an increasingly proficient buying process, which includes a decrease in the exchange cost, the cost of merchandise, process duration, and stock. Additionally, Min and Galle (1999) proposed operational advantages of buying rehearses through the use of web-based business. According to Croom (2000), the use of web and web-based business can give advantages, for example, the expanded productivity of request handling, a decreased expense due to JIT (Just-in-Time) stock administration and more prominent accessibility to give altered items and administrations.

Hill and Scudder (2002) featured a successful mix of electronic data exchange (EDI) and production network of the executives for improving efficiencies inside agro-nourishment supply chains and referenced use of various industry projects, for example, efficient purchaser reaction (EPR), estimated food records (EFR) and UCS II program to advance SCM and the use of EDI for playing out the normal exchanges. Scientists called EDI as an empowering technology to make ceaseless renewal conceivable to deal with the bullwhip impact and affirmed that bigger size firms may acquire proficiency from robotizing different exercises.

S. no.	Inventory management issues	References
1	EDI (Electronic Data Interchange) programs with vendors	Humphreys et al. (2001), Hill and Scudder (2002), Rahman (2003, 2004), Gunasekaran and Ngai (2004), Gurung (2006), Kumbhar (2011) and Rai et al. (2011)
2	JIT (Just-in-Time) delivery programs	Croom (2000), Rahman (2003, 2004), Folinas et al. (2006), Carr et al. (2008), Vanajakshi et al. (2009), Karmakar (2011) and Kumbhar (2011)
3	Communicate with vendors	Ghosh (1998), Rahman (2003, 2004), Schiefer (2003), Hobbs (2006), Aramyan et al. (2007), Deimel et al. (2008), Frentrup and Theuvsen (2009), Kumbhar (2011) and Chandraju et al. (2012)
4	Communicate out of stock	Ghosh (1998), Min and Galle (1999), Rahman (2003, 2004), Schiefer (2003), Gurung (2006), Carr and Kaynak (2007) and Kumbhar (2011)
5	Order ship date delays	Salin (1998), Closs and Kefeng (2000), Gunasekaran et al. (2001) and Rahman (2003, 2004)
6	Raw material inventory levels	Bookbinder and Dilts (1989), Kehoe and Boughton (2001), Batt (2003), Crouch (2003), Rahman (2003, 2004) and Fischer et al. (2007)
7	Emergencies affecting inventory	Simatupang and Sridharan (2002), Rahman (2003, 2004), Attaran (2007) and Rambaldi (2007)
8	Finished goods inventory levels	Simatupang and Sridharan (2002), Crouch (2003), Rahman (2003, 2004), Christopher (2005), Attaran (2007) and Kelepouris et al. (2007)
9	Field warehouses/depots inventory levels	Bookbinder and Dilts (1989), Mehta and Shah (2001), Rahman (2003, 2004), Gurung (2006) and Montanari (2008), Arora et al. (2022)
10	Field depots on out of stocks	Bookbinder and Dilts (1989), Kinsey (2000), Rahman (2003, 2004) and Gurung (2006)

Table 4.1. *Inventory management issues emerged from the literature*

Schiefer (2003) introduced certain IT advancement lines for advertise centered advances which can be used for agro-nourishment supply chains. These incorporate advanced coordination, multimedia collaboration, electronic correspondence organization, information entryway technology, virtual stages, specialist technology lead to EDI, ERP (Enterprise Resource Planning), multidimensional correspondence, electronic market systems, level and vertical information entries, web-based business exchanging stages, and segment advertise information and inventory network correspondence frameworks.

Rahman (2003, 2004) found that the greater part of the IT application in the production network was for stock administration. Web is used for the correspondence of stock-outs by clients to vendors or by organizations to their clients. The web empowered associations to rapidly organize electronic data interchange (EDI) with their clients. Further investigation through optional sources such as "The Sugar Technologists' Association of India (STAI)" and "Confederation of Indian Industry (CII)" revealed different issues identified with the stock administration of the sugar store network.

Based on all of these written works, the accompanying applicable issues related to stock administration exercises are shown in Table 4.1.

Based on the above arguments, the proposition was formulated that "*Are the identified issues impacted by using IT applications for managing inventory activities?*"

4.3. Research methodology

Qualitative and quantitative research approaches were used to identify and validate the inventory management issues in the sugar supply chain. Data was collected by visiting 30 sugar mills (four responses from each mill manager/department), 149 distributors (including brokers, wholesalers and retailers), 60 IT companies (only supply chain solution providers) and 11 LSSCC across Uttar Pradesh and Uttarakhand states in India for understanding the scope of IT applications to manage inventory activities. Interviews were conducted in both structured and unstructured ways with these experts. The survey was conducted, and respondents were asked to rate the issues on a five-point Likert scale (1 = Not at all relevant, 2 = Not relevant, 3 = Neutral, 4 = Relevant and 5 = Highly relevant) to check the scope of IT applications for managing the inventory activities.

4.4. Analysis

A total of 340 respondents were from sugar mills, distributors, IT companies and LSSCC.

Cronbach's alpha value (α) (Cho et al. 2019; Ferrara et al. 2019) of the instrument used for the identification of relevant issues in sugar industries inventory management activities was 0.862 (Hair et al. 1995).

4.4.1. *Factor analysis*

Factor analysis was carried out to identify relevant inventory management issues in the sugar supply chain. Factor analysis helped in identifying the valid factors and related measures (Clark and Bowles 2018; Carrizosa et al. 2020). The total of 340 responses received for establishing relevance of the issues is shown in Table 4.2.

Management of inventory activities using e-applications	Not at all relevant (1)		Not relevant (2)		Neutral (3)		Relevant (4)		Highly relevant (5)		Total respondents (N)
	Count	%	Count	%	Count	%	Count	%	Count	%	
I1	0	0.00	0	0.00	30	8.82	222	65.30	88	25.88	340
I2	0	0.00	0	0.00	25	7.35	166	48.82	149	43.82	340
I3	0	0.00	0	0.00	19	5.59	206	60.59	115	33.82	340
I4	0	0.00	0	0.00	22	6.47	222	65.29	96	28.24	340
I5	0	0.00	2	0.59	27	7.94	165	48.52	146	42.94	340
I6	0	0.00	0	0.00	48	14.12	214	62.94	78	22.94	340
I7	0	0.00	4	1.18	45	13.23	154	45.29	137	40.29	340
I8	0	0.00	0	0.00	40	11.76	119	35.00	181	53.24	340
I9	0	0.00	0	0.00	9	2.65	210	61.76	121	35.59	340
I10	0	0.00	0	0.00	20	5.88	191	56.18	129	37.94	340

Table 4.2. *Total number of responses received for establishing the relevance of the issues for managing inventory activities*

Descriptive statistics for various issues for managing inventory activities is presented in Table 4.3. The Pearson correlation with a two-tailed test of significance was used to test the statistical significance of the issues (I1–I10). Table 4.4 shows the correlation matrix for these 10 issues in inventory management activities. The table indicates that 44 out of the 45 correlations (97.78%) are significant at the 0.01 and 0.05 levels. There is no evidence of multicollinearity (Hair et al. 1995; Lüdtke et al. 2020). This provides a sufficient base to perform further analysis overall and for each issue with the help of factor analysis (Gupta and Gupta 2011; Carrizosa et al. 2020).

Notations	Issues	Mean	SD
I1	EDI (Electronic Data Interchange) programs with vendors	4.16	0.56
I2	JIT (Just-in-Time) delivery programs	4.36	0.61
I3	Communicate with vendors	4.28	0.55
I4	Communicate out of stock	4.21	0.54
I5	Order ship date delays	4.33	0.64
I6	Raw material inventory levels	4.08	0.60
I7	Emergencies affecting inventory	4.25	0.71
I8	Finished goods inventory levels	4.41	0.69
I9	Field warehouses/depots inventory levels	4.33	0.52
I10	Field depots on out of stocks	4.32	0.57

Table 4.3. *Descriptive statistics for issues to manage inventory activities*

Issues		I1	I2	I3	I4	I5	I6	I7	I8	I9	I10
I1	Pearson corr.	1									
	sig. (two-tailed)										
I2	Pearson corr.	.552**	1								
	sig. (two-tailed)	.000									
I3	Pearson corr.	.294**	.220**	1							
	sig. (two-tailed)	.000	.001								
I4	Pearson corr.	.243**	.299**	.625**	1						
	sig. (two-tailed)	.000	.000	.000							
I5	Pearson corr.	.204**	.167*	.476**	.567**	1					
	sig. (two-tailed)	.003	.016	.000	.000						
I6	Pearson corr.	.402**	.282**	.389**	.357**	.340**	1				
	sig. (two-tailed)	.000	.000	.000	.000	.000					
I7	Pearson corr.	.087	.355**	.428**	.383**	.376**	.345**	1			
	sig. (two-tailed)	.215	.000	.000	.000	.000	.000				
I8	Pearson corr.	.155*	.397**	.370**	.287**	.350**	.345**	.604**	1		
	sig. (two-tailed)	.026	.000	.000	.000	.000	.000	.000			
I9	Pearson corr.	.157*	.217**	.478**	.517**	.612**	.421**	.535**	.469**	1	
	sig. (two-tailed)	.024	.002	.000	.000	.000	.000	.000	.000		
I10	Pearson corr.	.281**	.232**	.483**	.517**	.651**	.408**	.475**	.370**	.844**	1
	sig. (two-tailed)	.000	.001	.000	.000	.000	.000	.000	.000	.000	

*Note: * Correlation is significant at the 0.05 level (two-tailed), and ** correlation is significant at the 0.01 level (two-tailed).*

Table 4.4. *Pearson correlation with significant two-tailed test for issues to manage inventory activities*

The overall significance level was tested using the KMO (Kaiser–Meyer–Olkin) and Bartlett's test. Table 4.5 indicates that the measure of sample adequacy is 0.801, which exceeds the minimum level of 0.45. The table shows that the value of the chi-square statistic is 1.001E3 at 45 degrees of freedom at a statistical significance level of 0.000, which is well below the minimum level of 0.05. This suggests that the construct is highly significant. In this construct, 10 issues, i.e. from I1 to I10, were reduced to three issues as follows (see Table 4.6 and Figure 4.1):

– vendor communications for out of stock, delays and depots inventory levels (FI1);

– emergencies affecting finished goods inventory (FI2);

– use of EDI and JIT for raw material inventory (FI3).

KMO measure of sampling adequacy	.801
Bartlett's test of sphericity	
Approx. chi-square	1.001E3
df	45
Sig.	.000

Table 4.5. *KMO (Kaiser–Meyer–Olkin) and Bartlett's test for issues to manage inventory activities*

Issues	**Component factors**			**Communalities**
	FI1	**FI2**	**FI3**	
I1	.190	-.094	**.895**	.845
I2	.013	.411	**.773**	.767
I3	**.684**	.173	.246	.558
I4	**.751**	.091	.233	.626
I5	**.809**	.144	.052	.678
I6	.439	.199	**.473**	.456
I7	.350	**.805**	.070	.775
I8	.214	**.842**	.176	.785
I9	**.782**	.403	.007	.774
I10	**.819**	.260	.102	.749
				Total
Sum of squares (eigen values)	3.366	1.862	1.786	**7.014**
Percentage of trace	33.657	18.623	17.857	**70.137**

Table 4.6. *VARIMAX-rotated component analysis factor matrix for issues to manage inventory activities*

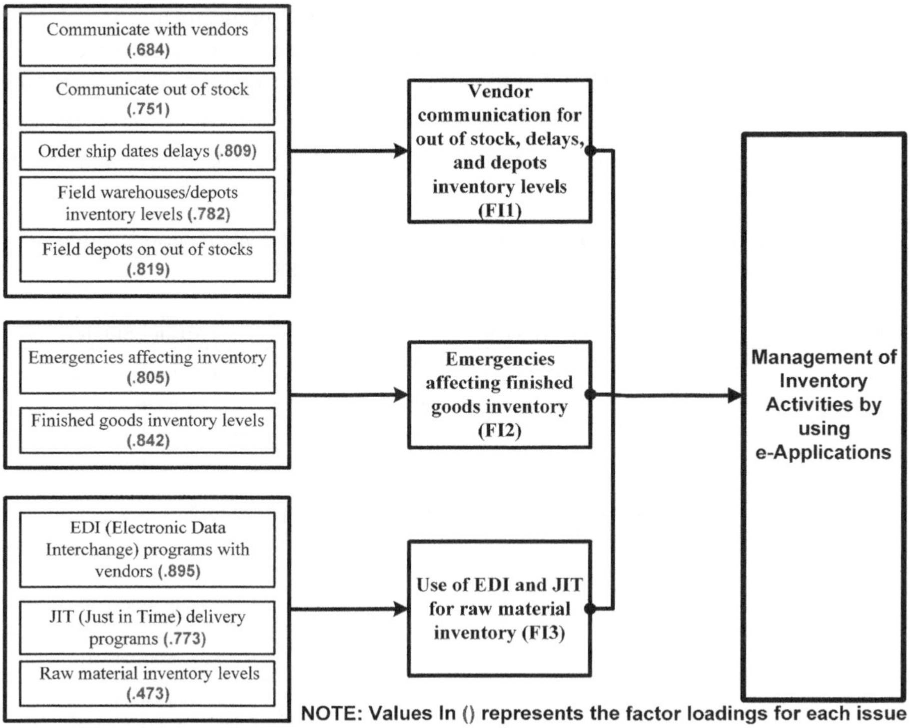

Figure 4.1. *A model for managing inventory activities using e-applications*

These issues were interpreted on the basis of a factor loading pattern. Table 4.6 shows that all the loadings calculated by incorporating VARIMAX Rotation are more than 0.30. The total percentage of variance for this construct is 70.137%, as obtained by adding all of the three issues or component factors. From Figure 4.1, it can be noted that the rotated factor solution for the issues I3, I4, I5, I9 and I10 loads significantly on FI1; I7 and I8 loads on FI2; I1, I2 and I6 loads on FI3. Therefore, it is likely that IT applications will be helpful in addressing these three issues (FI1, FI2 and FI3). A model is proposed for managing inventory activities using IT applications, as shown in Figure 4.1.

4.4.2. *Regression model*

The regression method was used to identify a relationship between various independent issues and a dependent issue (Du and Wang 2018; Ginkel 2019; Rights and Sterba 2020; Cremers et al. 2020; Kumar et al. (2021); Kashav et al. (2022)).

Descriptive statistics for reduced factors or issues for managing inventory activities is presented in Table 4.7. For inventory management activities, three issues were obtained, as shown in Figure 4.2. Of these, there is one dependent variable; the remaining two are independent variables:

– *dependent variable:* vendor communication for out of stock, delays and depots inventory levels (FI1);

– *independent variable:* (1) emergencies affecting finished goods inventory (FI2); (2) use of EDI and JIT for raw material inventory (FI3).

Notations	Issues	Mean	Std dev.	N
FI1	Vendor communication for out of stock, delays and depots inventory levels	4.29	0.463	340
FI2	Emergencies affecting finished goods inventory	4.33	0.630	340
FI3	Use of EDI and JIT for raw material inventory	4.20	0.462	340

Table 4.7. *Descriptive statistics for reduced factors or issues for managing inventory activities*

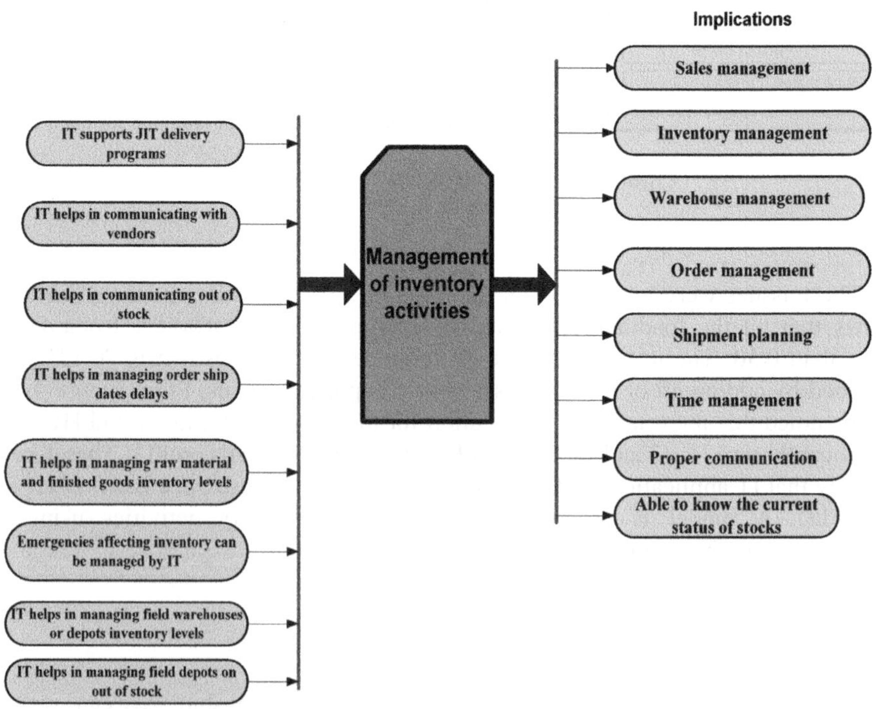

Figure 4.2. *Role of IT in inventory management. For a color version of this figure, see www.iste.co.uk/koubaa/optimization.zip*

The overall correlation coefficient is calculated between these three issues, i.e. FI1–FI3, as detailed in Table 4.8. These issues are highly correlated to each other, and no multicollinearity is detected. This indicates the suitable base for performing further analysis using the regression method. Table 4.9 shows the result of the analysis. The regression equation is:

$$FI1 = 1.773 + 0.328\,(FI2) + 0.263\,(FI3) \quad [4.1]$$

Issues	FI1	FI2	FI3
FI1	1		
FI2	0.553	1	
FI3	0.444	0.408	1

Table 4.8. *Correlations for reduced factors or issues for managing inventory activities*

Issues/predictors	Standardized coefficient (β)	T-test	Significance of T
(Constant)		6.980	**.000***
FI2	0.446	7.261	**.000***
FI3	0.262	4.277	**.000***
R^2 = 0.363; adjusted R^2 = 0.357; standard error of estimate = 0.3715; F-ratio = 57.814; significance = 0.000			

*Note: * Correlation is significant at the 0.05 level.*

Table 4.9. *Multiple regressions for vendor communication of inventory activities*

The results also indicate that the F-ratio for vendor communications (FI1) is 57.814 with a significant level of 0.000, which suggests a linear relationship between them. An R^2 of 0.363 and an adjusted value of R^2 of 0.357 have a small difference of 0.006. The standardized coefficients (β) for FI2 and FI3 are 0.446 and 0.262, respectively. Therefore, the results support this construct.

4.4.3. *Hypothesis testing*

The research hypothesis was tested using one-way ANOVA. This test was used for issues related to managing inventory activities. The hypothesis was stated as follows:

HYPOTHESIS 4.1.– there is no impact for using an IT application on communicating out of stock, emergencies for finished goods and raw material inventory for managing inventory activities.

Table 4.10 shows the results of one way ANOVA. The sum of squares value between groups (SS_{bg}) for FI2 is 30.665 with nine degrees of freedom (df), and the sum of squares within groups (SS_{wg}) is 50.724 with 196 degrees of freedom. Hence, the value of mean square between groups (MS_{bg}) is 30.665/9 = 3.407, and the mean square within groups (MS_{wg}) is 50.724/196 = 0.259. The value of F is 3.407/0.259 = 13.166 with 9 and 196 degrees of freedom, resulting in a probability (p) of 0.000. Similarly, the value of F can be calculated for FI3 as F (9, 196) = 17.627 at p = 0.000. The value of p is well below the minimum significance level of 0.05. Alternatively, the critical value of F can be calculated for FI2 with 9 and 196 degrees of freedom as 1.94 at p = 0.05. Table 4.10 shows that the calculated value of F (13.166) for FI2 is larger than the critical value of F. Hence, null hypothesis is rejected. In the regression equation (see equation [4.1]), the coefficients for FI2 and FI3 are different with 0.328 and 0.263, respectively. This results in accepting the alternate hypothesis that there are significant impacts of the use of IT systems on resolving these issues and therefore inventory activities can be managed by resolving these issues by implementing IT applications.

Issues		Sum of squares	df	Mean square	F	Sig.
FI2	Between groups	30.665	9	3.407	13.166	.000
	Within groups	50.724	196	.259		
	Total	81.388	205			
FI3	Between groups	19.606	9	2.178	17.627	.000
	Within groups	24.223	196	.124		
	Total	43.830	205			

Table 4.10. *Analysis of variance for issues to manage inventory activities*

4.5. Results and discussion

The results suggest that issues – vendor communication for out of stock, delays and depots inventory levels (FI1); emergencies affecting finished goods inventory (FI2); and use of EDI and JIT for raw material inventory (FI3) – are highly correlated. The results of hypothesis 4.1 test indicates that there is significant impact of using an IT application in communicating out of stock, emergencies for finished goods and raw material inventory for managing inventory activities. Therefore, implementing IT applications is likely to help facilitate the inventory management in the sugar supply chain.

Although the scope of this research is limited to finding out the impact of implementing the IT application in the inventory management of sugar industry, it would be appropriate to list by referring secondary sources. Some software that can be helpful in handling different inventory activities are as follows.

S. no.	Issues	IT solutions
1	– Tracking of shipment orders – Providing and updating stock status	– S2K inventory management software
2	– Managing the inventory stock levels – Reviewing the salesperson's monthly sales with customer identification	– S2K sales analysis software
3	– Managing the finished goods and raw material inventory – Managing emergencies affecting the inventory level – Providing information about the inventory levels	– RFID – Bar coding technique – SAP
4	– Field depots on out of stock – Field warehouses/depots inventory levels	– LOS – LIS
5	– JIT delivery of the raw materials and finished goods to fulfill the out-of-stock inventory	– EDI
6	– Managing order shipments – Managing order ship date delays	– LPS
7	– Communicating with vendors – Communicating out of stock	– Internet-based technologies such as e-mails, fax, etc.

Table 4.11. *IT solutions for managing inventory activities*

S2K inventory management software (SMC Data Systems n.d.) can be used for handling all the inventory processes such as tracking of shipment orders, providing and updating stock status; sales analysis software from the same producer can be helpful in managing the inventory stock levels and reviews the salesperson's monthly sales with customer identification. RFID and bar coding techniques (Attaran 2007) can be useful solutions for managing the finished goods and raw material inventory. By using these techniques, the current status of the stocks can be easily known and can therefore help mitigate emergencies, if any. A suitable integrated ERP system such as SAP-ERP can be implemented to integrate RFID and the bar coding technique and feeding the data into the computer system to provide required information about the inventory levels. Logistics operating systems (LOS) and logistics information systems (LIS) (Bookbinder and Dilts 1989; Gurung 2006) are the technology that can facilitate to manage the warehousing activities. EDI programs are necessary to share the information with vendors about the inventory level (Gurung 2006; Kumbhar 2011). This can facilitate the delivery of the raw materials and finished goods to replenish inventory. Logistics planning systems (LPS) can be used for the inventory management (Closs and Kefeng 2000) due to which order shipments can be managed. In the supply chain of the sugar industry, as the vendors and millers start using such IT applications, it can help prevent malpractices, if any. Table 4.11 shows the summary of IT solutions that can help manage different issues belonging to inventory activities.

4.6. Conclusion, limitations and scope for future work

From this examination, it may very well be presumed that implementing IT applications is likely to help encourage the stock administration in the sugar inventory network. There is a huge potential for the selection of most recent IT arrangements that can help achieve perceivability in the sugar industry production network framework. IT applications can assume an indispensable job in overseeing field stops, stockroom, crude material and completed great stock levels, request shipping delays, etc. Without the legitimate administration of stock exercises, the store network arrangement of any industry such as sugar is likely to affect its creation and benefit. Accordingly, for consistent business exercises, it is advisable to make a typical foundation of the IT application.

The scope of this study is limited to the conditions of Uttar Pradesh and Uttarakhand in India. Respondents from different states in India can be included for future investigations. The after-effects of this investigation bring an open door for investigating the effect of using IT applications in the administration of various sorts of exercises, for example, vendor the board, transportation and conveyance exercises. This should be possible by including various inventory network accomplices. Then, certain IT arrangements can be proposed for dealing with these exercises. The use of IT applications can lead to changes in the sugar inventory network in Uttar Pradesh and beyond. Through further research, an online business design can be created by indicating the value of IT at various levels. The sugar industry can build its intensity over other different enterprises by using IT applications.

4.7. References

Aramyan, L.H., Lansink, A.G.J.M.O., Vorst, J.G.A.J.V., Kooten, O.V. (2007). Performance measurement in agri-food supply chains: A case study. *Supply Chain Management: An International Journal*, 12(4), 304–315. doi: 10.1108/13598540710759826.

Arora, M., Kumar, R., Raju, T.B. (2022). Identification of issues in the cold chain of Indian frozen food. *International Journal of Logistics Economics and Globalisation (IJLEG)*. doi: 10.1504/IJLEG.2023.10049129.

Attaran, M. (2007). RFID: An enabler of supply chain operations. *Supply Chain Management: An International Journal*, 12(4), 249–257.

Bangwal, D., Suyal, J., Kumar, R. (2022). Hotel building design, employee health and performance in response to COVID 19. *International Journal of Hospitality Management*, 103, 103212.

Batt, P.J. (2003). Building trust between growers and market agents. *Supply Chain Management: An International Journal*, 8(1), 65–78. doi: 10.1108/13598540310463378.

Blanchard, D. (2010). *Supply Chain Management Best Practices*, 2nd edition. John Wiley & Sons Inc, Hoboken, NJ.

Bookbinder, J.H. and Dilts, D.M., (1989). Logistics information systems in a just-in-time environment. *Journal of Business Logistics*, 10(1), 50–67.

Burton, L.J. and Mazerolle, S.M. (2011). Survey instrument validity part I: Principles of survey instrument development and validation in athletic training education research. *Athletic Training Education Journal*, 6(1), 27–35.

Carr, A.S. and Kaynak, H. (2007). Communication methods, information sharing, supplier development and performance: An empirical study of their relationships. *The International Journal of Operations and Production Management*, 27(4), 346–370. doi: 10.1108/01443570710736958.

Carr, A.S., Kaynak, H., Muthusamy, S.K. (2008). The cross-functional coordination between operations, marketing, purchasing and engineering and the impact on performance. *The International Journal of Manufacturing Technology and Management*, 13(1), 55–76.

Carrizosa, E., Guerrero, V., Morales, D.R., Satorra, A. (2020). Enhancing interpretability in factor analysis by means of mathematical optimization. *Multivariate Behavioral Research*, 55(5), 748–762. doi: 10.1080/00273171.2019.1677208.

Chamarbagwala, R. and Sharma, G. (2010). Industrial de-licensing, trade liberalization, and skill upgrading in India [Online]. Available at: http://bengal.missouri.edu/~sharmag/Aug2010_Chamarbagwalla_Sharma_skills.

Chandraju, S., Raviprasad, B., Kumar, C.S.C. (2012). Studies on the implementation of system application product (SAP) materials management (MM Module) for financial support in sugar cane cultivation. *IOSR Journal of Business and Management*, 3(5), 7–12 [Online]. Available at: www.iosrjournals.org.

Cho, S.J., Shen, J., Naveiras, M. (2019). Multilevel reliability measures of latent scores within an item response theory framework. *Multivariate Behavioral Research*, 54(6), 856–881. doi: 10.1080/00273171.2019.1596780.

Christopher, M. (2005). *Logistics and Supply Chain Management: Creating Value-Adding Networks*. Prentice Hall Financial Times, Harlow.

Clark, D.A. and Bowles, R.P. (2018) Model fit and item factor analysis: Overfactoring, underfactoring, and a program to guide interpretation. *Multivariate Behavioral Research*, 53(4), 544–558. doi: 10.1080/00273171.2018.1461058.

Closs, D.J. and Kefeng, X. (2000). Logistics information technology practice in manufacturing and merchandising firms: An international benchmarking study versus world class logistics firms. *International Journal of Physical Distribution & Logistics Management*, 30(10), 869–886.

Cremers, J., Pennings, H.J.M., Ley, C. (2020). Regression models for cylindrical data in psychology. *Multivariate Behavioral Research*, 55(6), 910–925. doi: 10.1080/00273171.2019.1693332.

Croom, S. (2000). The impact of web-based procurement on the management of operating resources supply. *The Journal of Supply Chain Management*, 36(1), 4–13. doi: 10.1111/j.1745-493X.2000.tb00065.x.

Crouch, D. (2003). Online purchasing revisited. *CMA Management*, 77(4), 37–39.

Deimel, M., Frentrup, M., Theuvsen, L. (2008). Transparency in food supply chains: Empirical results from German pig and dairy production. *Journal on Chain and Network Science*, 8(1), 21–32.

Desai, N. (2008). IT industry in India. [Online]. Available at: http://www.studymode.com/essays/Industry-India-158052.html.

Du, H. and Wang, L. (2018) Reliabilities of intraindividual variability indicators with autocorrelated longitudinal data: Implications for longitudinal study designs. *Multivariate Behavioral Research*, 53(4), 502–520. doi: 10.1080/00273171.2018.1457939.

Fatemeh, E.S., Ramzani, U., Ebrahimzadeh, S., Sargolzae, S., Sepasgozar, S. (2020). Technology acceptance in e-governance: A case of a finance organization. *Journal of Risk and Financial Management*, 13(7), 138. doi: 10.3390/jrfm13070138.

Ferrara, C., Martella, F., Vichi, M. (2019). Probabilistic disjoint principal component analysis. *Multivariate Behavioral Research*, 54(1), 47–61. doi: 10.1080/00273171.2018.1485006.

Fischer, C., Gonzalez, M.A., Henchion, M., Leat, P. (2007). Factors influencing trust supporting mechanisms in European agri-food chains. *Food Economy*, 4(1), 40–49.

Floyd, F.J. and Widaman, K.F. (1995). Factor analysis in the development and refinement of clinical assessment instruments. *Psychological Assessment*, 7(3), 286–299.

Folinas, D., Manikas, I., Manos, B. (2006). Traceability data management for food chains. *British Food Journal*, 108(8), 622–633. doi: 10.1108/00070700610682319.

Frentrup, M. and Theuvsen, L. (2009). Information technologies and transparency in agri-food supply chains: Empirical results from the German pig and dairy production. *Proceedings of the 7th EFITA Conference*. Academic Publishers, Wageningen, 655–666 [Online]. Available at: http://www.efita.net/.

Ghosh, A. (1998). Private internet service providers in India. [Online]. Available at: http://www.india50.com/isp.html [Accessed 5 January 2012].

Ginkel, J.R.V. (2019). Significance tests and estimates for R2 for multiple regression in multiply imputed datasets: A cautionary note on earlier findings, and alternative solutions. *Multivariate Behavioral Research*, 54(4), 514–529. doi: 10.1080/00273171.2018.1540967.

Goldman Sachs (2000). Report on IT global services. [Online]. Available at: www.goldman-sachs.com.

Goodridge, E. (2000). Behind the numbers: E-business grows amid market turmoil. Information Week No. 811 [Online]. Available at: http://www.informationweek.com/bizint/biz811/turmoil.htm.

Goswami, R., Juneja, E., Sharma, S. (2005). Agribusiness sector in rural India and increasing opportunities of e-commerce. *Proceedings of Conference on Marketing to Rural Consumers*. Indian Institute of Management Kozhikode, 145–148 [Online]. Available at: http://dspace.iimk.ac.in/bitstream/2259/524/1/RM70.pdf.

Gunasekaran, A. and Ngai, E.W.T. (2004). Information systems in supply chain integration and management. *European Journal of Operational Research*, 159(2), 269–295. doi: 10.1016/j.ejor.2003.08.016.

Gunasekaran, A., Patel, C., Tirtiroglu, E. (2001). Performance measures and metrics in a supply chain environment. *International Journal of Operations & Production Management*, 21(1/2), 71–87. doi: 10.1108/01443570110358468.

Gupta, G. and Basole, A. (2020). India's information technology industry: Prospects for growth and role in structural transformation. *Decision*, 47, 341–361. doi: 10.1007/s40622-020-00269-z.

Gupta, S.L. and Gupta, H. (2011). *SPSS 17.0 for Researchers*. International Book House Pvt. Ltd, New Delhi.

Gurung, A. (2006). A survey of information technologies in logistics management. In *Proceedings of the 37th Annual Conference of SWDSI*. The University of Mississippi Press, Bricktown, Oklahoma City, 618–626.

Hair, J.F., Anderson, R.E., Tatham, R.L., Black, W.C. (1995). *Multivariate Data Analysis: With Readings*, 4th edition. Prentice Hall, New Jersey.

Harwood, C.J. (2001). Electronic business revolution: Opportunities and challenges in the 21st century. *Kybernetes*, 30(1), 103–115.

Heckmann, P., Shorten, D., Engel, H. (2003). *Supply Chain Management at 21: The Hard Road to Adulthood*. Booz Allen Hamilton Inc., New York.

Hill, C.A. and Scudder, G.D. (2002). The use of electronic data interchange for supply chain coordination in the food industry. *Journal of Operations Management*, 20(4), 375–387. doi: 10.1016/S0272-6963(02)00017-7.

Hobbs, J.E. (2006). Liability and traceability in agri-food supply chains. In *Quantifying the Agri-Food Supply Chain*, Ondersteijn, C.J.M., Wijnand, J.H.M., Huirne, R.B.M., van Kooten, O. (eds). Springer, Dordrecht, 85–100.

Humphreys, P., Lai, M., Sculli, D. (2001). An inter-organisational information system for supply chain management. *International Journal of Production Economics*, 70(3), 245–255. doi: 10.1016/S0925-5273(00)00070-0.

Jacoby, D. (2009). *Guide to Supply Chain Management: How Getting it Right Boosts Corporate Performance*, 1st edition. The Economist Books, Bloomberg Press, New York.

Kalakota, R. and Robinson, M. (1999). *e-Business: Roadmap for Success*. Addison-Wesley, Reading, MA.

Kansara, S., Modgil, S., Kumar, R. (2022). Structural transformation of fuzzy analytical hierarchy process: A relevant case for Covid-19. *Operations Management Research*, 16, 450–465.

Karmakar, K.G. (2011). Organised agri-food retailing in India. National Bank for Agriculture and Rural Development (NABARD), Mumbai. [Online]. Available at: http://www.iimidr.ac.in/iimi/media/pdf/Case/Management%20Case%20-%20Samridhdhhi.pdf.

Kashav, V., Garg, C.P., Kumar, R., Sharma, A. (2022). Management and analysis of barriers in the maritime supply chains (MSCs) of containerized freight under fuzzy environment. *Research in Transportation Business & Management*. doi: 10.1016/j.rtbm.2022.100793.

Kehoe, D.F. and Boughton, N.J. (2001). New paradigms in planning and control across manufacturing supply chains – The utilisation of internet technologies. *International Journal of Operations & Production Management*, 21(5/6), 582–593. doi: 10.1108/01443570110390345.

Kelepouris, T., Pramatari, K., Doukidis, G. (2007). RFID-enabled traceability in the food supply chain. *Industrial Management & Data Systems*, 107(2), 183–200. doi: 10.1108/02635570710723804.

Kinsey, J. (2000). Electronic systems in the food industry: Entropy, speed and sales. *Proceedings of E-Business Transformation: Sector Developments and Policy Implications BRIE Conference*. Brookings Institution, Washington, DC.

Kumar, R. (2020). E-applications for managing trans-logistics activities in sugar supply chain in North India. *International Journal of Asian Business and Information Management*, 11(1), 15. doi: 10.4018/IJABIM.2020010106.

Kumar, R., Organero, M.M., Agrawal, R. (2010). XML secure documents for a secure e-commerce architecture. *Global Journal of Enterprise Information System*, 2(1), 35–45.

Kumar, R., Agrawal, R., Sharma, V. (2013). e-applications in Indian agro-food supply chain: Relationship among enablers. *Global Business Review*, 14(4), 711–727.

Kumar, R., Agrawal, R., Sharma, V. (2015). IT enablement in sugar supply chain: An approach for farmers. *International Journal of Business Performance and Supply Chain Modelling*, 7(4), 360–381.

Kumar, R., Gupta, P., Gupta, R. (2021). A TISM and MICMAC analysis of factors during COVID-19 pandemic in Indian Apparel Supply Chain. *IJISSCM – IGI Global*, 15(1), 24. doi: 10.4018/IJISSCM.287133.

Kumbhar, R.D. (2011). ERP system for effective management of co-operative sugar industries: A case study of Sahyadri SSK Ltd Shiravade Karad (M.S.). *International Journal of Information Technology and Knowledge Management*, 4(1), 33–37.

Lüdtke, O., Robitzsch, A., West, S.G. (2020). Analysis of interactions and nonlinear effects with missing data: A factored regression modeling approach using maximum likelihood estimation. *Multivariate Behavioral Research*, 55(3), 361–381. doi: 10.1080/00273171.2019.1640104.

Mehta, K.T. and Shah, V. (2001). E-commerce: The next global frontier for small businesses. *Journal of Applied Business Research*, 17(1), 87–94 [Online]. Available at: http://www.journals.cluteonline.com/index.php/JABR/article/view/2067/2254.

Min, H. and Galle, W. (1999), Electronic commerce usage in business-to-business purchasing. *International Journal of Operations & Production Management*, 19(9), 909–921. doi: 10.1108/01443579910280232.

Moni, M. (2003), Impact of economic reforms on Indian agricultural sector: Application of geomatics technology to reduce marginalization and vulnerability of small farmers in India [Online]. Available at: http://www.gisdevelopment.net/application/agriculture/production/agric0003pf.htm [Accessed 24 December 2011].

Montanari, R. (2008). Cold chain tracking: A managerial perspective. *Trends in Food Science & Technology*, 19(8), 425–431. doi: 10.1016/j.tifs.2008.03.009.

Palevich, R.F. (1999). Supply chain management. *Hospital Material Management Quarterly*, 20(3), 54–63.

Rahman, Z. (2003). Internet-based supply chain management: Using the internet to revolutionize your business. *International Journal of Information Management*, 23, 493–505. doi: 10.1016/j.ijinfomgt.2003.09.004.

Rahman, Z. (2004). Use of internet in supply chain management: A study of Indian companies. *Industrial Management & Data Systems*, 104(1), 31–41. doi: 10.1108/02635570410514070.

Rai, S., Sutirtha, C., Suprateek, S. (2011). ICT innovation in contemporary India: Three emerging narratives. *ASCI Journal of Management*, 41(1), 53–64 [Online]. Available at: http://journal.asci.org.in.

Rambaldi S. (2007). L'informatica al servizio dell'alimentare. *Largo Consumo*, 86–88 [Online]. Available at: http://www.largoconsumo.info/.

Rights, J.D. and Sterba, S.K. (2020). New recommendations on the use of R-squared differences in multilevel model comparisons. *Multivariate Behavioral Research*, 55(4), 568–599. doi: 10.1080/00273171.2019.1660605.

Ross, D.F. (1997). *Competing Through Supply Chain Management: Creating Market-Winning Strategies Through Supply Chain Partnerships*. Chapman & Hall, New York.

Salin, V. (1998). Information technology in agri-food supply chains. *International Food and Agribusiness Management Review*, 1(3), 329–334. doi: 10.1016/S1096-7508(99)80003-2.

Schiefer, G. (2003). New technologies and their impact on agriculture, environment and the food industry. *Proceedings of the European Federation for Information Technology in Agriculture (EFITA) Conference*, Debrecen, 3–11.

Simatupang, T.M. and Sridharan, R. (2002). The collaborative supply chain. *The International Journal of Logistics Management*, 13(1), 15–30. doi: 10.1108/0957409021 0806333.

SMC Data Systems (n.d.). Distribution management software solutions: Providing a global view [Online]. Available at: http://www.smcdata.com/software-choices/distribution-management-software-solutions-1.html [Accessed 22 December 2011].

Timmers, P. (1999). *Electronic Commerce: Strategies and Models for Business-To-Business Trading*. John Wiley & Sons, Singapore.

Vanajakshi, L., Subramanian, S.C., Sivanandan, R. (2009). Travel time prediction under heterogeneous traffic conditions using global positioning system data from buses. *Intelligent Transport Systems*, 3(1), 1–9. doi: 10.1049/iet-its:20080013.

5

Dynamics of the Rice Supply Chain System in Punjab, Pakistan

Rice is considered the second most important cereal crop (after wheat) in Pakistan. Out of the total production, 40–45% is consumed domestically and the rest (up to 50%) is exported to different countries, especially Gulf countries. The extant exploratory research intends to understand the value chain dynamics and market margins of rice in domestic markets to explore the barriers and opportunities for the development of a sustainable rice value chain system. After developing the pretested semi-structured questionnaire, primary data were collected from 90 value chain actors of the main rice-producing regions of the Punjab: Lahore (Nankana Sahib), Muzaffargarh and Gujranwala. Moreover, focus group interviews were conducted with purposively selected 20 key actors of the chain-like farmers, small-scale contractors, millers, wholesalers and retailers. The data collected from these actors were thus analyzed using descriptive statistics and margin analysis to analyze the value chain system of rice and then present true insight. Findings prioritized the major problems faced by stakeholders, for example, lack of appropriate extension services, infrastructure and an inadequate marketing information system were ranked as major factors in creating the vulnerability at the upstream level of the chain, and commission agents/wholesalers were seen getting the major share in marketing margin of rice, compared to other actors. The study explored that our marketing system critically lacks the adoption of advanced production technologies and skilled labor as it may help reduce the qualitative and quantitative losses of the rice industry. Furthermore, stakeholders emphasized the importance of providing input subsidies and building a network of micro-institutions to help rural smallholder farmers to invest in their crops sustainably.

Chapter written by Muhammad Bilal AHSAN, Abd-Ur-REHMAN, Zohaib JABBAR and Hafiz Zahid MAHMOOD.

5.1. Introduction

Pakistan has a total population of 235 million, with nearly 2/3 of them living in rural areas. The agricultural sector, including grain crops, livestock and horticultural products, is a major driver of Pakistan's economy. The agricultural sector of Pakistan contributes almost 21% to gross domestic product (GDP) as well as 29.9% to gross national product (GOP 2022). Agriculture plays a pivotal role in the economy, from the primary source of import substitution to exports of valuable goods.

In order to get from the farmer to the final consumer, agricultural commodities are passed through many hands. There is a complex nexus between all of these players, and each one contributes something to the product's overall marketing value. Prices of agricultural commodities in traditional supply chains are volatile, depending upon the number of participants involved as well as factors such as product shelf life and country of origin. The agricultural value chains fail to implement the fundamentals of the marketing concept, namely, the classification of products based on market segmentation. The first step in deciding who or what to target is brainstorming to identify relevant groups of people to choose from. Positioning is crucial in producing segment-specific products based on the needs of consumers in traditional and high-end markets as well as in reaching the most profitable consumer groups (Annie 2002; Kotler 2003).

Rice (*Oryza sativa*) and its secondary products are relied upon by over half of the world's population for a significant portion of their daily caloric intake, particularly in Asia, where over 200 billion people get anywhere from 60% to 70% of their food from rice and rice-based foods (FAO 2014). Cultivating rice is a fascinating part of many traditional and religious celebrations all over the world, and it is also a major step toward food self-sufficiency and food security in many low-income developing countries (Lundy et al. 2004; Gebremeskel 2010). It provides billions of households with the majority of their annual income. Several Asian and African nations rely heavily on rice exports to generate foreign exchange and tax revenue.

Pakistan's rice exports account for 2.4% of the total value added in agriculture, and 0.5% in GDP. In Pakistan, rice is the second most important staple crop after wheat. Rice is the third most planted crop in Pakistan, after wheat and cotton, according to the area under cultivation. More than 10% of Pakistan's arable land is devoted to this crop. It is the third most produced agricultural commodity in the world, after maize and sugarcane, and is a major part of food for half of the global population. Rice accounts for 20% of India's total crop yield, while Brazil is the largest non-Asian grower, followed by the USA (Indiamart 2009). After China, India and Indonesia, Pakistan is the fourth most populous country in the world (FAO 2014). Among the many types of basmati rice grown in Pakistan, Super

Kernel Basmati, Pk-385 and super basmati varieties, as well as the pk-198 and pk-198 Basmati are the most prominent ones (Giraud 2008).

Table 5.1 provides a supplemental classification of the aforementioned rice products and sub-products according to the production stages at which they are created. In order to properly map and identify the processes involved in the value chain, it is necessary to categorize these products.

Level	Product type
Primary	Rice including husk (rough/paddy)
Semi-processed	(Brown rice) husked rice
Processed	Milled or broken rice

Table 5.1. *Classification of rice products (source: Pakistan Bureau of Statistics, Government of Pakistan, 2015–2016)*

The initial stage of the rice value chain is the harvesting of paddy rice, also known as rice in a husk. To create brown or husked rice, the outer, inedible keel is removed during the processing stage. Milled rice is rice that has been refined by removing the husk and the bran coating. Rice that has been milled but imperfect is called broken or wrecked rice.

Punjab ranked first in rice production, contributing 58% of the total rice production of the country, followed by Sindh (with 29% contribution), KPK (with 10% contribution) and Balochistan (with 3% contribution), as detailed in Table 5.2. In Pakistan, Basmati rice occupies 52% of the country's total farmland. Approximately 45–50% of rice yield is consumed domestically (locally), while the rest is exported (GOP 2015–2016).

Variety	Pakistan	Punjab	Sindh	KPK	Balochistan
	Area (000)	%	%	%	%
Total	2,748.5	69.27	20.96	6.03	3.47
Basmati	1,638.2	100	-	-	-
IRRI	780.1	21.76	58.54	-	17.69
Others	330.2	5,445	2,641	18.14	-
Total	6,811.4	58	29	10	3
Basmati	3,392.1	100	-	-	-
IRRI	2,598.2	21.50	58.51	-	15.99
Others	799.1	52.85	31.03	17.12	-

Table 5.2. *Shares of provinces in area and production of rice ("000" hectares, %) (source: Agricultural Statistics; Government of Pakistan, 2019)*

Pakistan's third largest export commodity is rice. Pakistan exports over half of its rice crop, making it the third largest rice exporter in the world. Exported rice typically comes from two main varieties: basmati and IRRI. Among Pakistan's major exports, fragrant basmati rice accounts for 25%, and basmati accounts for 10% (2 million tons). The exporters of rice are all independent businesspeople; the government plays almost no role. Pakistan's major rice trading rivals are India, Thailand and Vietnam. Global rice exports have fallen from US$ 2.035 billion (2014–2015) to US$ 1.860 billion in 2015–2016. The amount of rice in the hands of traders has increased by 2.0 million metric tons (GOP 2015–2016).

5.1.1. *Rice supply chain*

There is a common perception that provisions under the sanitary and phytosanitary agreement of the World Trade Organization may distort the competitiveness of agricultural products from developing countries (Mustafa and Ahmad 2003). Consumers in both international and domestic (local) markets are increasingly becoming conscious about product quality (Mendoza 1995). Rice is grown and processed by employing strategies to reach markets at the international level as consumer-driven value chains across borders. However, the extant research will prioritize the problems faced by rice value chain actors and explore sustainable solutions.

Rice has made a gradual transition from a cereal crop to a cash crop in some regions. This is because its popularity in both domestic and international markets keeps rising. Reasons for adequate production of rice include an ideal climate for growing paddy, availability of suitable land and a shift in agricultural focus away from cash crops toward food crops. With its rising production and consumption, rice has the potential to cause large fluctuations in Pakistan's rural economic situation. Rice marketing in our country is still disorganized despite the attention of government and international development agencies (Kelly and Mather 2012). In the conventional system of marketing, farmers first produce their goods, and then these products are pushed toward buyers. Many farmers have no direct contact with the end-user of their products and limited influence over the expected profit or cost of their inputs (RIU 2010).

5.1.2. *Objectives*

1) To map the major actors and core activities in rice value chain.

2) To estimate the percentage marketing margins for each actor in the rice supply chain system.

3) To explore/prioritize problems and recommend suitable interventions for the development of rice value chain.

5.2. Methodology

The basis for this study was the major rice-producing regions, commonly called the rice belt of Pakistan, which is situated in Doabs (a land between two rivers). The main rice-producing regions were selected as study areas, and a representative sample (N=100) was randomly collected. For this purpose, two regions: Gujranwala, Muzaffargarh and Lahore (Nankana sahib) were selected with equal share of each margin. The irrigation source is non-perennial and perennial canals, and the rest of the water need is accomplished through groundwater pumping. The main crops of these regions are rice and wheat. Moreover, Bajra, Mash, Moong, Jawar, sunflower and mustard are also being cultivated to some extent. In this study, marketing margins were calculated using the following formula:

$$\text{PS} = \frac{Px}{-Pr}$$

where PS is the producers' share, P_x as the price paid by producers and P_r as the retail price of rice. Market margin analysis reveals the consumer's value for the dollar or rupee they pay, actually represents the marketing services provided by intermediaries as well as shows the efficiency of the marketing system. Profitability analysis was calculated using the following formula (Ajala and Adesehinwa 2008):

$$Gross\ Profit = V - C\ or\ PQ - \Sigma\ Pi\ qi$$

where P = price per Kg, C = total production cost, Pi = input price, Q = total production, Qi = input quantity, V = production value.

Net profit percentage of margin = Net margin/gross marketing margin * 100

Net profit percentage of sales price = Net margin/average sales price * 100

Overall, 20 farmers, 10 contractors, 10 rice millers, 20 wholesalers and 30 retailers were part of the total sample. Farmers were asked about the data of their cost, access to market and availability of quality inputs. Data were analyzed through descriptive statistics and margin analysis to understand the dynamics of the rice supply chain system.

5.3. Results and discussion

This section is divided into three sections based on the objectives of the study. The first section includes the value chain mapping of rice focusing on major stakeholders in the chain and exploring the key issues and gaps in the marketing system of rice. In the second section, the authors analyzed and compared the marketing margins of all actors in the domestic marketing system of rice by adopting the margin analysis approach. The third section focuses on constraints, opportunities and recommendations based on the qualitative data of most ranked constructs by actors and consumers of premium quality and traditionally processed rice.

5.3.1. *Value chain mapping of rice*

The market/value chain mapping approach was used to depict the known marketing channels of rice in the traditional supply chain. Problems and gaps in the chain are highlighted in red, as shown in Figure 5.1. Each actor may have persistent or on-spot relations with other actors in the chain and distinguish the quality attributes based on the quality perceived by the end consumer.

Due to the country's antiquated and decade-old market system, producers cannot rely on retailers or distributors for the introduction of innovative agricultural post-harvest practices. Rice growers typically transport the raw products to various middlemen and commission agents. Similarly, farmers, particularly small-scale farmers dealing with rice, are obligated to sell their produce to specific wholesalers (aarhtis), rather than open markets in order to raise money to cover their production expenses. The low prices they receive for their crops make it difficult for them to invest in advanced production technologies.

Commission agents in the grain markets deal with a wide variety of products, from the highest to the lowest quality (known locally as A, B and C, respectively) due to the dynamic nature of the market. However, farmers' profit margins tend to be lower when they sell their goods at auction due to issues like contract enforcement, inadequate information and weak bargaining position. There are some commission agents who also supply major agricultural products in order to distort the market equilibrium. They are the main players/drivers because of the pivotal role they play as intermediaries in the chain. On the contrary, today's premium quality value chains are driven by supermarkets and department stores in the metropolitan cities, where consumer preferences have a direct impact on the standards that are applied in the field.

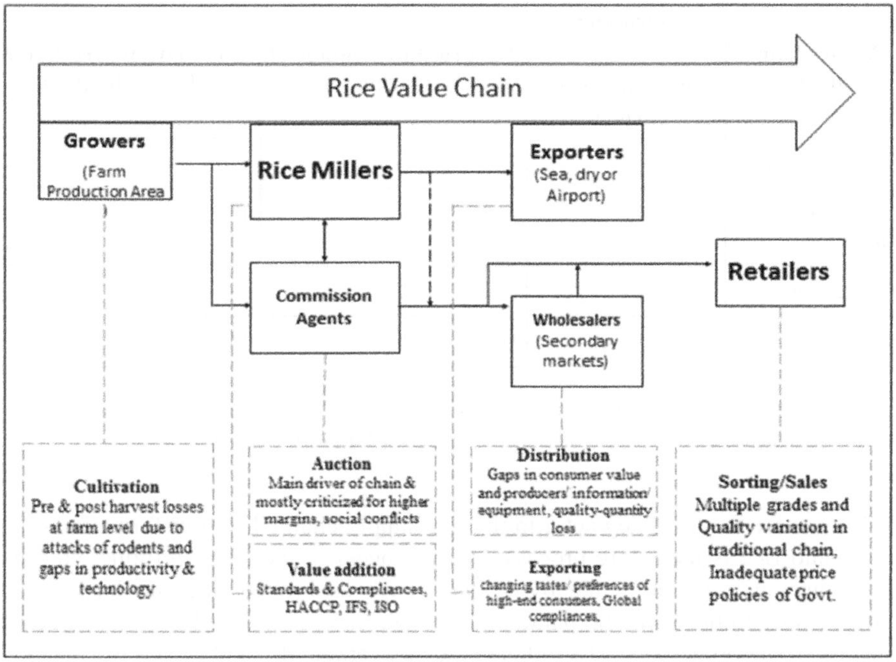

Figure 5.1. *Value chain system of rice in Punjab, Pakistan (note: gaps are highlighted in red). For a color version of this figure, see www.iste.co.uk/koubaa/optimization.zip*

5.3.2. *Margin analysis of the marketing system*

Intermediaries/contractors play a pivotal role in the rice industry after farmers. Produce is purchased by the contractor, who then transports it to the rice mills for processing. There are two distinct kinds of agricultural contractors: those who buy crops as they are being grown and those who buy them after they have been harvested. Contractors in Pakistan may not have a lot of formal education, but they still make decent money. The above calculation indicates that contractors pay farmers 1,175 rupees per maund when purchasing the crop. These operations incurred an average cost of 92.67, transported it to the rice mills and then sold the rice for 1,350 rupees. Remarkably, the average gross marketing margin was 175 rupees; this is quite adequate for the modest business practices of the industry and the ensuing 12.97% marketing margin. A contractor's net profit was 47.04% (or 82.33 rupees), based on their sales. The percentage of net profit on sales was 6.10 rupees.

Value chain actors	Marketing margin (percentage of price)	Net profit percentage of margin	Net margin percentage of the sales price
Rice contractors	12.97	47.04	6.10
Millers (rice mills)	63.06	26.08	17.24
Wholesalers	1.95	86.2	1.68
Retailers	18.57	80.24	14.90

Table 5.3. *Percentage distribution (share in the marketing margin of all actors) (source: author's estimations from primary data)*

Millers are the third key actor in the rice supply chain. The transformation of the paddy into rice is their core process. Many processes, including refining, coloring, processing and separating the by-products such as broken rice and husk oil are performed by millers. The preparation of rice for sale as finished good is the central function of the processors. The millers paid an average of 1,350 rupees per maund to the contractors to purchase the paddy and an average of 3,655 rupees per maund to sell the rice. The total processing, advertising and distribution expenses are 3,727 Pakistani rupees. In comparison to the margins of other actors in the marketing chain, the average gross marketing margin was 2,305 rupees or 63.06%. The millers had a net margin of 630 rupees or 26.80%. After all expenses were deducted, the net profit as a percentage of the selling price was 17.24%.

The central player that links all other actors in the marketing chain is the wholesaler. Due to the connections wholesalers maintain with both domestic and international retailers, they serve as a pivotal component in the economic wheel. Some wholesalers have direct connections with global rice traders. Profit margins for wholesalers in Pakistan are higher than other actors in the rice supply chain. Based on the data presented above, wholesalers paid an average of 3,655 rupees per maund to millers for rice and get 3,727.5 rupees per ton when they resold it. The overall marketing costs averaged out to just 10 rupees, which is negligible. The net marketing margins amounted to 1.95% of sales or 72.5 rupees. The net profit margin for wholesalers averaged 86.20% or 62.5 rupees. Lastly, the median percentage of the selling price was 1.68%.

Distribution of major problems faced by farmers	**Frequency**	**Percentage**
Limited area under cultivation	4	20
Lack of knowledge and skills	5	25
Traditional cultivation methods	2	10
Lack of advanced technology	3	15
Lack of financial resources	2	10
Outdated technologies	3	15
Lack of technical support	1	5
Distribution of major problems faced by rice millers		
Higher input cost	2	10
Untrained labor	2	20
Public policies	2	20
Credit and cash flow problems	1	10
Load shedding and prices of utilities	2	20
Distribution of major problems faced by wholesalers		
Dynamic markets	2	10
Losses during loading and unloading	5	25
Lack of specialized labor	4	20
Poor infrastructure	3	15
Deferred payments	2	10
Lack of credit for expansion	2	10
Distribution of major problems faced by retailers		
Poor quality and multiple grades	9	30
High transportation cost	7	23.3
Price differentiation	5	16.6
Lack of attractive packaging and branding	6	20
Non-availability of storage facilities	3	10

Table 5.4. *Ranking of problems faced by supply chain actors. (source: author's calculations from primary data)*

In this study, retailers are considered to be playing the most crucial role in the rice value chain. There is not a single store that does not store the products according to the need of the surrounding areas. Rice is purchased by retailers from both wholesalers and millers. As per the average price paid by retailers, the wholesale price of rice was 3,727.5 rupees and the retail price was 4,577.5 rupees.

The average marketing expense borne by the retailers was 168 Indian rupees. The average gross marketing margin was 850 rupees or 18.57%. The average net margin for all stores was 80.24% or 682 rupees. The final average percentage of the selling price was 14.90%.

5.3.3. *Challenges and opportunities for the rice industry of Pakistan*

High paddy prices and high marketing costs were found to be affecting the premium quality production in the rice supply chain of Punjab, Pakistan. Almost 20% of respondents attributed the high price of paddy to the high cost of labor. Pakistan has significantly greater production costs than either India or China. It is also true that our country's marketing costs were much higher than those of its international rivals (Dawe et al. 2008; Beltran et al. 2016). Lower economies of scale and outdated technologies of rice mills, as well as high packaging and transport costs, all contribute to the increasing marketing costs of rice in Pakistan, which in turn raise the cost of working capital.

Constraints	**Recommendations/opportunities**
Producers (production)	
Low yield of paddy (rice)	Use of advanced production technologies and labor-intensive machinery
High production costs of labor, and material inputs (seeds, pesticides and fertilizer)	In order to reduce costs and increase revenues, direct marketing and sustainable use of resources according to the requirement of soil (experts' opinion)
Climate change, which causes significant qualitative and quantitative losses	Access to climate-smart technologies, such as crop diversification, the cultivation of varieties adapted to different ecosystems, carefully managed irrigation and the use of renewable energy-powered machinery
Farmers are not satisfied with high-quality varieties in the market	Government and private sector need to focus through an integrated approach to bridge the gap. Farmers' access to high-quality seeds should be improved
Prevalent attacks of pests and diseases	Disease and pest-resistant varieties should be developed and good agricultural practices as well as integrated pest management and organic pesticides (bio-control agents) should be promoted
Regulatory issues in accessing the credit	Presence of agricultural credit through the Agricultural Credit Policy to help farmers should be easy to get and less time consuming
Commission agents and contractors (paddy in primary markets)	
Paddy quality is degraded due to a lack of access to modern storage and mechanical drying facilities	Available information on the recommended varieties, together with their agronomic traits, grain quality features and resistance to pests and diseases

Limited access to modern storage and mechanical drying facilities; thus, depend mostly on sun drying that reduce the quality of paddy	Public–private partnerships should be enhanced to set up grain trading posts with full post-harvest and market facilities, with the goal of serving the unique needs of individual farmers, farmer cooperatives and other value chain actors, with a focus on the most populous surplus provinces. Insist that farmers' groups and merchants use less of the inefficient traditional solar drying method that causes significant losses in the drying process and moisture meters should be provided to assess quality
Transportation costs are high because of fuel prices of diesel, labor and maintenance	To combat this, the government should initiate the farm-to-market support system: (1) decrease transportation costs for farm inputs and outputs; (2) reduce post-harvest losses; (3) increase access to basic services; and (4) link farms
Limited compliance in labeling the different quality grades for raw paddy and some developed standards have little advocacy for disseminating these to new traders	Government should approve the National Level Standards for grain grading classification of paddy and milled rice; it specifies what kinds of packaging, labels and prices consumers can expect to see in the market, as well as the quality and safety standards that those products must adhere to
Rice millers (processing)	
Lack of yield-enhancing and cost-reducing technologies, low milling recovery, drying and storage of paddy	Development of yield-enhancing and cost-efficient technologies and practices to increase yield and reduce cost
High cost of electricity and diesel also contributes to high milling and mechanical drying costs	Rice mills that generate rice husks can be used for power generation, reducing the cost of milling and mechanical drying instead of electricity and diesel
Retailers (marketing)	
High price, low quality of milled rice, disillusion of consumers by multiple grades	The availability of storage facilities for appropriate drying and storage of paddy improve grain quality and milling efficiency, mitigating the effects of high procurement costs, low milled rice quality, and consumer disappointment caused by multiple grades, improve the enforcement of grain grading and pricing standards to maintain quality, especially in peak harvest months
Congested grain markets during peak season leads to inefficiency	Proper facilities should be built to improve traditional storage systems so that produce could be handled efficiently

Table 5.5. *Challenges and opportunities for rice value chain actors (source: author's notes form the value chain actors survey)*

The reasons for high domestic prices of rice include low yields, high production, and marketing costs, number of players and a lack of investment in market infrastructure such as transportation, storage, drying and milling facilities. The supply chains of rice as a whole were impacted by these issues. It is important to determine what kinds of innovative actions and investments can make the rice industry more emerging, globally competitive and sustainable in light of the change.

5.4. References

Ajala, M.K. and Adesehinwa, A.O.K. (2008). Analysis of pig marketing in Zango Kataf local government area of Kaduna State, Nigeria. *Tropicultura*, 26(4), 229–239.

Annie, I.M. (2002). Marketing strategies of Dawlance and LG. Department of Management Sciences, Isra University Hyderabad, 10 June.

Beltran, J.C., Bordey, F.H., Moya, P.F., Launio, C.C., Manalili, R.G., Marciano, E.B., San Valentin, M.R., D. Valencia, M.S., Dawe, D.C (2016). Rice prices and marketing margins. In *Competitiveness of Philippine Rice in Asia*, Bordey, F.H., Moya, P.F., Beltran, J.C., Dawe, D.C. (eds). International Rice Research Institute, Philippine Rice Research Institute, Maligaya, Science City of Muñoz, Nueva Ecija, 55–73.

Dawe, D.C., Moya, P.F., Casiwan, C.B., Cabling, J.M. (2008). Rice marketing systems in the Philippines and Thailand: Do large numbers of competitive traders ensure good performance? *Food Policy*, 33, 455–463.

FAO (2014). FAOSTAT, Food and Agriculture Organization of the United Nations (FAO). 1 January.

FAO (2015). FAOSTAT, Food and Agriculture Organization of the United Nations (FAO). Crop Prospects and Food Situation, 3 October. Quarterly Global Report [Online]. Available at: https://www.fao.org/markets-and-trade/publications/detail/en/c/1446277/.

FAO (2016). FAOSTAT, Food and Agriculture Organization of the United Nations (FAO). 3 October.

Gebremeskel, B.K. (2010). Rice value chain in Metema District, North Gondar, Ethiopia: Challenges and opportunities for innovation. Dissertation, Haramaya, 116.

Giraud, G. (2008). Range and limit of geographical indication scheme: The case of basmati rice from Punjab, Pakistan. *International Food and Agribusiness Management Review*, 11(1), 51–76.

GOP (Government of Pakistan) (2014–2015). "Economic Survey", Economic Advisor's Wing, Finance Division, Ministry of Finance, Islamabad.

GOP (Government of Pakistan) (2015–2016). "Economic Survey", Economic Advisor's Wing, Finance Division, Ministry of Finance, Islamabad.

GOP (Government of Pakistan) (2022). "Economic Survey", Economic Advisor's Wing, Finance Division, Ministry of Finance, Islamabad.

Indiamart (2009). World rice production [Online]. Available at: http://sourcing.indiamart.com/agriculture [Accessed 9 May 2012].

Kabuje, F.L. (2008). Analysis of the Value Chain for Hides and Skin in Tanzania: The case of Dodoma and Arusha Regions. Doctoral Dissertation, Morogoro, 109.

Kelly, V.A. and Mather, D. (2012). Farmers' production and marketing response to rice price increases and fertilizer subsidies in the Office Du Niger. Food Security International Development Working Paper 146941, Michigan State University, Department of Agricultural, Food, and Resource Economics. doi: 10.22004/ag.econ.146941.

Kotler, P.D. (2003). *Strategic Marketing for Nonprofit Organizations*. Prentice Hall, Upper Saddle River, NJ.

Lundy, M., Hellin, J., Meijer, M. (2004). Farmer organization, collective action and market access in Meso-America. *Food Policy*, 34(1), 16–22.

Mendoza, E.G. (1995). The terms of trade, the real exchange rate, and economic fluctuations. *International Economic Review*, 36(1), 101–137.

Mustafa, K. and Ahmad, S. (2003). Barriers against agricultural exports from Pakistan: The role of WTO sanitary and phytosanitary agreement. *The Pakistan Development Review*, 42(4), 487–510.

RIU (2010). Value chain approach to poverty reduction and development of livelihood [Online]. Available at: http://www.researchintouse.com [Accessed 25 June 2011].

6

Solving the Multi-Objective Four-Dimensional Transportation Problem using the Strength Pareto Evolutionary Algorithm

This research deals with the four-dimensional (4D) multi-objective transportation problem in the General Company for Food Stuff Trading because of its great role in managing the transportation process. The transportation costs, time and distance were obtained for the different demand centers, the different supply centers, the distribution line and the different means of transportation. The multi-objective 4D transportation model was solved using the strength Pareto evolutionary algorithm and using the coding formula via the MATLAB program, the lowest transportation cost, transportation time and distance were obtained.

The goal of the researcher was to use new tools, which are as follows:

1) Using a realistic applied case on the 4D transmission method with multiple objectives using AI algorithms in an integrated manner.

2) Using the decoding procedure to speed up the program execution process within the calculator.

3) Using the Pareto group to find out the extent to which the objectives of the 4D model are homogeneous with each other.

Chapter written by Zainab Alaa HAMEED and Taicir Moalla LOUKIL.

6.1. Introduction

Haley (1962) proposed a procedure for solving the 3D transportation problem, and he noted that many objectives must be considered and optimized at the same time; and in this case, the problem becomes a multi-objective 3D transportation problem. In general, in the transportation problem, the distances of tracks or roads are neglected, i.e. they are not taken into account between the different supply (source) centers and the different demand centers (requesting parties) (Halder et al. 2017). It is interesting, in the transportation system, to add the distance of tracks or roads along with the different means of transportation. In this case, the 3D transport problem is transformed into a 4D transport problem, and when more than one criterion is considered for this problem, it becomes a multi-objective 4D transportation problem (Sahoo et al. 2020). Halder et al. (2017) proposed a formulation and solved a special kind of the 4D multi-element transportation problem, with constant charge and profits for alternative and breakable elements in crisp and fuzzy environments (Halder et al. 2019). Bera et al. (2018) presented a paper for formulating and solving the problem of 4D multi-element transport with budget constraints using repeated environments.

In this chapter, we propose a mathematical model of the multi-objective 4D transport problem and try adapted SPEA2 algorithms to eliminate potential weaknesses of the model and achieve good results. In addition, the proposed algorithm provides a good performance in terms of convergence and diversity.

In higher dimensional objective spaces, the number of non-dominated solutions increases rapidly. This presents a significantly greater challenge, for example, for the archiving strategies and in general for the algorithms to keep the desired convergence properties, while maintaining a good distribution of solutions. In fact, this observation was the reason why great importance was placed on accurate density estimation when designing the strength Pareto evolutionary algorithm.

The research aims to build a mathematical model for the 4D multi-objective transportation problem for the purpose of reducing the costs of transporting different homogeneous or heterogeneous materials from different places due to their presence of multiple destinations requesting the use of different means of transportation to achieve a set of goals, including cost, time and distance.

The chapter is organized as follows: in section 6.2, a mathematical formulation of the 4D multi-objective transportation problem is presented. Section 6.3 gives a brief description of the strength Pareto evolutionary algorithm metaheuristic. In section 6.4, a real case problem is solved using the SPEA2 metaheuristic. Section 6.5 concludes the chapter and gives some perspectives.

6.2. A mathematical model for the multi-objective four-dimensional transportation problem

The multi-objective 4D transportation problem can be described as follows:

1) The transfer process is between several different sources (supply) and several different demand centers.

2) Different means of transportation are considered, according to the type of transported materials, whether they are homogeneous or heterogeneous.

3) There is more than one distribution line that enables the decision-maker to achieve their goals, including reducing (cost–time–distance), and it is possible to take a number of other objectives, for example, the cost of satisfying the customer, the level of inventory (maintaining the lowest level of storage), and expansion by opening other branches.

The mathematical model of the multi-objective 4D transport problem is as follows (Anderson et al. 1997; Bakhayt 2016; Bera et al. 2018; Halder et al. 2019; Sahoo et al. 2020):

$$Min \quad Z_1 = \sum_{i=1}^{m} \sum_{j=1}^{n} \sum_{k=1}^{p} \sum_{q=1}^{d} C_{i,j,k,q} * X_{i,j,k,q} \qquad [6.1]$$

$$Min \quad Z_2 = \sum_{i=1}^{m} \sum_{j=1}^{n} \sum_{k=1}^{p} \sum_{q=1}^{d} T_{i,j,k,q} * X_{i,j,k,q} \qquad [6.2]$$

$$Min \quad Z_3 = \sum_{i=1}^{m} \sum_{j=1}^{n} \sum_{k=1}^{p} \sum_{q=1}^{d} S_{i,j,k,q} * X_{i,j,k,q} \qquad [6.3]$$

The y constraints include the record of the different means of transport and the different centers of demand and supply, in addition to the transport of different types of materials.

Subject to:

$$\sum_{q=1}^{d} \sum_{j=1}^{n} \sum_{k=1}^{p} X_{i,j,k,q} = a(i) \qquad \forall\ i = 1, \dots, m \qquad [6.4]$$

$$\sum_{i=1}^{m} \sum_{k=1}^{p} \sum_{q=1}^{d} X_{i,j,k,q} = b(j) \qquad \forall\ j = 1, \dots, n \qquad [6.5]$$

$$\sum_{i=1}^{m} \sum_{j=1}^{n} \sum_{k=1}^{p} X_{i,j,k,q} = d(q) \qquad \forall\ q = 1, \dots, d \qquad [6.6]$$

$$\sum_{i=1}^{m} \sum_{j=1}^{n} \sum_{q=1}^{d} X_{i,j,k,q} = L(k) \qquad \forall\ k = 1, \dots, p \qquad [6.7]$$

$$\sum_{i=1}^{m} a(i) = \sum_{j=1}^{n} b(j) = \sum_{q=1}^{d} d(q) \qquad [6.8]$$

$$X_{i,j,k,q} \geq 0 \qquad [6.9]$$

knowing that:

m: the number of display sources $(i = 1,2, \ldots, m)$;

n: the number of demand centers $(j = 1,2, \ldots, n)$;

p: the number of types of materials $(k = 1,2, \ldots, p)$;

d: the number of different types and modes of transportation $(q = 1,2, \ldots, d)$;

$X_{i,j,k,q}$: the quantity of materials transported of the commodity of type (k) from source (i) to demand center (j). Transfer of type (q);

$C_{i,j,k,q}$: the cost of transporting one unit of materials (k) from the source (i) to the demand center (j) with the transportation of type (q);

$T_{i,j,k,q}$: the time to transfer one unit of material (k) from source (i) to demand center (j) with transfer type (q);

$S_{i,j,k,q}$: the distance traveled to transport one unit of materials (k) from the source (i) to the demand center (j) a type of transport (q);

$a(i)$: the amount of material available in the source (supply) (i);

$b(j)$: the quantity of materials needed by the order center (j);

$d(q)$: the capacity of the transport medium (q) used in the transport process;

$L_{(K)}$: the quantity of material (k).

Equation [6.4] indicates that the sum of the quantities transported from different sources (supply) for each type of different materials is equal to the equipment available in these sources.

Equation [6.5] indicates that the total quantities transferred from different sources (supply) to the different requesting parties must meet the needs of these requests and for each of the different types of materials.

Equation [6.6] indicates that the total quantities transported from different sources (supply) must be transferred to the different requesting parties y the appropriate transportation means for the material type k.

Equation [6.7] indicates that the sum of the quantities of type (q) transferred from different sources (supply) to the different requesting parties y all transportation means of each type of material.

Equation [6.8] indicates that the quantities of the various commodities received from all the different demand sides are equal to the quantities of commodities supplied from all different sources and the total quantities of the various commodities supplied from all the different sources.

Equation [6.9] indicates non-negativity constraints where variables must be equal to or greater than zero.

This model is difficult to solve accurately in an acceptable time. Therefore, we proposed to solve it using an approximate method: the metaheuristic strength Pareto evolutionary algorithm.

RESEARCH CONTRIBUTIONS.–

Using a realistic applied case on the 4D transport method with multiple goals; using the Pareto group to find out the extent to which the objectives of the 4D model are homogeneous with each other.

6.3. Strength Pareto evolutionary algorithm (SPEA2)

SPEA2 is a population-based metaheuristic. It uses a normal population and an archive (outer set), starting with the initial population and an empty archive, the following steps are executed for each iteration. Firstly, all non-dominated populations are copied to the archive, and any individuals or duplicates (regarding target values) are removed from the archive during the update process. If the size of the updated archive exceeds a predetermined limit, more archive members are deleted using a pooling technique that preserves the characteristics of the non-dominated front. Next, the efficiency values are assigned to each of the population and archive members (Zitzler and Thiele 1998; Zitzler et al. 2001):

– Each individual i in the archive is assigned an estimated value of S(i) ∈ [0,1), which at the same time represents the appropriate fitness value of F(i); S(i) represents the number of members of community j that are dominated by or equal to i objective values, divided by population size plus one.

– The fitness F(j) for individual j in the population is calculated by summing the estimated values of S(i) for all members of archive i that dominate or equal to j and adding one at the end (Kim et al. 2004).

Pseudo code of SPEA2 (Costa et al. 2017)

1) Generate initial population P_0 and empty archive A_0. Set t=0.

2) Calculate fitness values of individuals in P_t and A_t.

3) A_{t+1} =non-dominated individuals in P_t and A_t.

If the size of $A_{t+1} > N$, then reduce A_{t+1}

If the size of $A_{t+1} < N$, then fill A_{t+1} with dominated individuals in P_t and A_t

4) If t >T, then output the non-dominated set A_{t+1}.

Stop.

5) Fill mating pool by binary tournament selection with replacement on A_{t+1}.

6) Apply recombination and mutation operators to the mating pool and set P_{t+1} to the resulting population. Set t=t+1 and go to step 2.

6.4. Application of SPEA2 to a real case problem

The General Company for Foodstuff Trading was established in 1997 and aims to contribute to supporting the national economy in the economic/commercial fields in accordance with development plans and planning decisions. Its activity is based on importing and marketing foodstuffs within its marketing and commercial plans. The data were obtained from the General Company for Food Stuff Trading for (8) demand centers, (8) supply centers, (2) distribution lines and (4) various transportation means.

The use of AI algorithms is appropriate in complex and large-scale problems where it is difficult to model and solve using exact methods, and given the number of variables in a multi-objective 4D transportation mode. Then, we will use an AI algorithm, namely the *SPEA2*, to solve the problem. However, before solving, we must find a decoding procedure that helps the algorithm find good solutions in record time. The steps of the decoding procedure to solve the problem of the multi-objective 4D transport model are as follows (Mukherjee et al. 2021):

Procedure: decoding

Input

m: the number of supplier centers;

n: the number of demand centers;

P: the number of items;

d: the number of vehicles.

The quantities of items available in the supplier centers i; $a(i)$:

b(j): the demand of items in the demand center j;

the number of various materials in the supply sources; $L(k)$:

the capacity of vehicle q; $d(q)$:

$C_{i,j,k,q}$ = the cost of transportation a unit of item k for the supplier center i to the demand center j by vehicle

$\forall\, q = 1, \dots, d \quad i = 1, \dots, m \quad k = 1, \dots, p \quad j = 1, \dots, n$

the particle for transportation. $V(i|j|q|k)$:

Output

the amount transportation $x_{i,j,q,k}$:

Step 1: $x_{i,j,k,q} \leftarrow 0 \;\forall\, q \in m \quad i \in n \quad k \in p \quad j \in d$

Step 2: $i^* \leftarrow argmin\,\{C_{i,j,q,k} | V(i) \neq 0 \; i \in n\}$ select a i lowest cost

$j^* \leftarrow argmin\{C_{i,j,q,k} | V(j) \neq 0 \; j \in d\}$ select a j lowest cost

$k^* \leftarrow argmin\{C_{i,j,q,k} | V(k) \neq 0 \; k \in p\}$ select a k lowest cost

$q^* \leftarrow argmin\{C_{i,j,q,k} | V(q) \neq 0 \; q \in m\}$ select a q lowest cost

Step 3: $x_{i^*,j^*,q^*,k^*} \leftarrow \min\{a_{q^*}, b_{i^*}, L_{K^*}, d_{j^*}\}$ update the availabilities

$$a_{q^*} = a_{q^*} - x_{i^*,\,j^*,q^*,\,k^*} \quad b_{i^*} = b_{i^*} - x_{i^*,\,j^*,q^*,\,k^*} \quad L_{K^*} = L_{K^*} - x_{i^*,\,j^*,q^*,\,k^*} \quad d_{j^*} = d_{j^*} - x_{i^*,\,j^*,q^*,\,k^*}$$

Step 4: if $a_{q^*} = 0, then \;\; V(i^*|j|q|k) \leftarrow 0$

$if \; b_{i^*} = 0 \;\; then \;\; V(i|j^*|q|k) \leftarrow 0$

$if \; L_{K^*} = 0 \;\; then \;\; V(i|j|q^*|k) \leftarrow 0$

$if \; d_{j^*} = 0 \;\; then \;\; V(i|j|q|k^*) \leftarrow 0$

Step 4: $if \; V(i|j|q|k) = 0 \;\forall q \in m \quad i \in n \quad k \in p \quad j \in d \;\; then \; output \; x_{i,j,q,k}$

Else return to step 2

Data of the problem

	demand node 1	demand node 2	demand node 3	demand node 4	demand node 5	demand node 6	demand node 7	demand node 8	Supply
supplier 1									2000
supplier 2									2000
supplier 3									2000
supplier 4									2000
supplier 5									2000
supplier 6									2000
supplier 7									2000
supplier 8									2000
demand	1800	1600	2000	2500	2200	1700	2100	2100	16000

Table 6.1. *Supply and demand quantities*

1,1								
	demand node 1	demand node 2	demand node 3	demand node 4	demand node 5	demand node 6	demand node 7	demand node 8
supplier 1	831	842	919	929	867	773	739	834
supplier 2	876	723	752	976	704	893	918	812
supplier 3	928	772	985	813	925	712	824	758
supplier 4	739	740	886	846	949	936	719	705
supplier 5	771	806	730	936	832	968	712	813
supplier 6	998	813	939	989	746	966	830	884
supplier 7	745	781	863	982	819	944	714	945
supplier 8	738	821	830	708	763	765	941	816
	demand node 1	demand node 2	demand node 3	demand node 4	demand node 5	demand node 6	demand node 7	demand node 8
supplier 1	41	39	65	30	52	53	76	33
supplier 2	54	59	65	29	61	44	29	23
supplier 3	65	78	26	60	22	45	76	66
supplier 4	45	77	61	74	69	31	68	60
supplier 5	46	47	48	51	65	35	55	63
supplier 6	27	34	32	62	27	21	46	59
supplier 7	21	66	26	29	52	76	35	45
supplier 8	37	66	70	78	39	59	65	43
	demand node 1	demand node 2	demand node 3	demand node 4	demand node 5	demand node 6	demand node 7	demand node 8
supplier 1	11	13	20	9	22	21	12	24
supplier 2	22	16	17	9	16	20	20	6
supplier 3	22	25	23	11	25	7	23	17
supplier 4	16	20	25	7	23	22	10	8
supplier 5	10	25	24	20	12	5	7	22
supplier 6	19	9	13	20	16	13	9	8
supplier 7	9	16	5	16	12	20	12	15
supplier 8	14	6	16	12	18	21	11	25

2,1								
	demand node 1	demand node 2	demand node 3	demand node 4	demand node 5	demand node 6	demand node 7	demand node 8
supplier 1	847	763	759	841	961	811	839	807
supplier 2	721	855	732	822	904	927	997	776
supplier 3	794	965	879	721	992	897	819	828
supplier 4	930	705	933	822	761	902	876	944
supplier 5	932	921	783	965	722	739	738	881
supplier 6	731	959	717	750	790	880	706	899
supplier 7	958	862	934	746	911	989	932	838
supplier 8	919	934	751	722	940	935	722	800
	demand node 1	demand node 2	demand node 3	demand node 4	demand node 5	demand node 6	demand node 7	demand node 8
supplier 1	69	47	56	71	42	43	57	80
supplier 2	39	23	43	64	25	58	55	21
supplier 3	69	72	28	54	59	21	52	52
supplier 4	68	58	21	30	31	75	36	25
supplier 5	71	41	45	78	22	68	35	68
supplier 6	50	80	31	36	64	65	47	80
supplier 7	58	33	64	76	41	69	33	24
supplier 8	78	59	42	33	60	43	69	77
	demand node 1	demand node 2	demand node 3	demand node 4	demand node 5	demand node 6	demand node 7	demand node 8
supplier 1	12	25	21	22	19	22	14	6
supplier 2	5	14	22	11	10	12	25	15
supplier 3	9	24	17	13	14	14	17	15
supplier 4	13	5	21	23	13	17	22	6
supplier 5	12	17	11	17	17	19	10	23
supplier 6	9	21	9	25	21	20	18	23
supplier 7	24	9	11	9	7	20	17	14
supplier 8	19	24	17	22	22	13	25	21

1,2								
	demand node 1	demand node 2	demand node 3	demand node 4	demand node 5	demand node 6	demand node 7	demand node 8
supplier 1	741	975	999	761	847	749	994	923
supplier 2	945	771	890	847	735	785	744	893
supplier 3	936	997	774	738	736	861	917	979
supplier 4	960	921	861	843	855	827	991	902
supplier 5	896	756	762	721	775	795	891	954
supplier 6	984	904	771	823	974	785	804	850
supplier 7	700	702	937	750	849	961	883	789
supplier 8	783	903	908	977	776	979	844	832
	demand node 1	demand node 2	demand node 3	demand node 4	demand node 5	demand node 6	demand node 7	demand node 8
supplier 1	21	33	80	38	55	74	62	64
supplier 2	61	31	77	61	51	47	20	33
supplier 3	67	22	61	52	25	45	71	36
supplier 4	52	26	80	45	63	33	76	61
supplier 5	74	57	66	56	80	27	67	49
supplier 6	74	77	40	65	41	38	22	58
supplier 7	58	41	60	55	79	64	43	34
supplier 8	28	45	34	53	41	67	62	30
	demand node 1	demand node 2	demand node 3	demand node 4	demand node 5	demand node 6	demand node 7	demand node 8
supplier 1	8	12	25	20	18	25	19	9
supplier 2	18	21	18	15	23	20	15	18
supplier 3	10	25	23	18	25	18	6	6
supplier 4	14	8	13	23	18	15	23	10
supplier 5	22	9	18	9	7	10	11	10
supplier 6	9	19	25	13	5	25	9	23
supplier 7	11	12	16	25	17	16	7	14
supplier 8	15	25	24	13	16	5	11	20

2,2								
	demand node 1	demand node 2	demand node 3	demand node 4	demand node 5	demand node 6	demand node 7	demand node 8
supplier 1	882	965	937	799	906	867	751	808
supplier 2	878	860	916	896	978	716	757	944
supplier 3	70	74	32	46	75	70	40	39
supplier 4	66	26	51	80	63	44	57	71
supplier 5	77	22	75	69	57	65	41	65
supplier 6	26	53	58	49	40	70	66	78
supplier 7	31	67	26	74	77	39	45	21
supplier 8	26	39	43	28	27	53	50	41
	demand node 1	demand node 2	demand node 3	demand node 4	demand node 5	demand node 6	demand node 7	demand node 8
supplier 1	55	71	20	52	62	38	69	53
supplier 2	42	61	67	35	67	62	20	76
supplier 3	33	28	66	75	50	43	20	38
supplier 4	33	72	45	66	45	54	25	40
supplier 5	51	32	23	74	57	74	35	72
supplier 6	46	57	55	24	72	71	21	40
supplier 7	65	53	30	31	60	74	45	28
supplier 8	24	29	64	64	51	77	40	50
	demand node 1	demand node 2	demand node 3	demand node 4	demand node 5	demand node 6	demand node 7	demand node 8
supplier 1	17	18	13	6	20	5	5	23
supplier 2	21	9	7	9	15	22	11	6
supplier 3	7	17	14	14	9	18	8	22
supplier 4	25	17	14	25	13	16	24	16
supplier 5	22	17	16	21	8	18	6	19
supplier 6	6	14	21	14	20	20	17	9
supplier 7	14	5	19	12	12	6	18	16
supplier 8	11	15	23	6	24	23	18	19

1,3								
	demand node 1	demand node 2	demand node 3	demand node 4	demand node 5	demand node 6	demand node 7	demand node 8
supplier 1	992	703	871	853	862	999	822	700
supplier 2	922	916	940	996	714	903	788	777
supplier 3	932	787	821	743	958	920	976	705
supplier 4	873	716	842	806	705	976	953	743
supplier 5	724	834	962	828	976	752	755	863
supplier 6	991	704	946	962	743	745	869	787
supplier 7	766	966	808	744	719	840	808	805
supplier 8	765	714	879	885	906	774	825	921
	demand node 1	demand node 2	demand node 3	demand node 4	demand node 5	demand node 6	demand node 7	demand node 8
supplier 1	34	34	55	78	57	27	25	61
supplier 2	63	63	42	22	69	36	76	74
supplier 3	58	72	69	79	21	35	44	23
supplier 4	56	37	50	31	25	40	22	38
supplier 5	60	64	49	60	79	29	40	22
supplier 6	22	28	73	55	59	41	64	31
supplier 7	41	71	41	61	34	27	68	63
supplier 8	47	28	47	42	44	73	53	64
	demand node 1	demand node 2	demand node 3	demand node 4	demand node 5	demand node 6	demand node 7	demand node 8
supplier 1	25	24	11	22	14	7	24	11
supplier 2	14	15	8	16	10	7	15	25
supplier 3	6	25	22	17	19	9	18	12
supplier 4	6	6	21	11	23	8	23	11
supplier 5	18	9	10	19	24	17	16	7
supplier 6	21	21	9	21	20	6	10	24
supplier 7	19	24	11	14	10	22	25	7
supplier 8	12	21	22	14	19	20	5	11

2,3								
	demand node 1	demand node 2	demand node 3	demand node 4	demand node 5	demand node 6	demand node 7	demand node 8
supplier 1	837	801	852	872	886	855	737	949
supplier 2	875	860	873	824	930	856	869	988
supplier 3	807	811	985	747	990	773	746	761
supplier 4	997	889	848	958	754	821	935	704
supplier 5	875	891	752	867	873	724	701	822
supplier 6	932	975	968	769	933	782	860	738
supplier 7	963	825	701	835	797	962	841	988
supplier 8	723	850	732	703	700	772	937	820
	demand node 1	demand node 2	demand node 3	demand node 4	demand node 5	demand node 6	demand node 7	demand node 8
supplier 1	73	63	59	28	74	47	56	60
supplier 2	55	71	38	21	56	59	73	76
supplier 3	24	46	28	54	73	51	76	69
supplier 4	76	48	49	38	77	42	60	49
supplier 5	68	54	42	77	53	77	32	66
supplier 6	37	36	68	79	64	70	59	45
supplier 7	53	65	67	37	55	71	24	79
supplier 8	80	50	60	68	21	42	44	80
	demand node 1	demand node 2	demand node 3	demand node 4	demand node 5	demand node 6	demand node 7	demand node 8
supplier 1	23	22	5	7	20	25	13	16
supplier 2	15	12	19	21	18	25	10	25
supplier 3	17	22	17	6	15	7	20	10
supplier 4	17	10	7	9	11	14	23	10
supplier 5	19	17	21	10	18	18	20	24
supplier 6	5	24	17	7	7	11	13	6
supplier 7	16	6	6	23	8	20	24	11
supplier 8	5	6	6	19	5	16	10	17

1,4								
	demand node 1	demand node 2	demand node 3	demand node 4	demand node 5	demand node 6	demand node 7	demand node 8
supplier 1	780	840	709	717	832	936	811	748
supplier 2	789	746	895	710	855	855	749	825
supplier 3	878	732	789	770	848	736	846	966
supplier 4	809	983	767	791	719	933	990	960
supplier 5	724	746	890	828	979	725	995	882
supplier 6	775	705	739	921	778	754	901	801
supplier 7	846	922	772	768	828	734	930	907
supplier 8	718	865	734	740	749	853	740	731
	demand node 1	demand node 2	demand node 3	demand node 4	demand node 5	demand node 6	demand node 7	demand node 8
supplier 1	72	56	37	74	39	46	54	70
supplier 2	43	29	61	70	47	27	44	28
supplier 3	47	74	60	43	63	69	44	23
supplier 4	35	47	27	50	73	39	51	25
supplier 5	67	32	44	62	63	35	60	29
supplier 6	73	74	36	70	21	40	78	39
supplier 7	75	66	63	57	61	42	64	38
supplier 8	54	73	37	55	46	53	44	20
	demand node 1	demand node 2	demand node 3	demand node 4	demand node 5	demand node 6	demand node 7	demand node 8
supplier 1	9	7	15	19	8	22	17	11
supplier 2	18	15	25	6	5	21	9	21
supplier 3	21	22	9	13	10	24	15	12
supplier 4	15	24	7	11	16	7	25	6
supplier 5	18	15	11	11	23	15	15	15
supplier 6	21	10	13	7	5	13	19	12
supplier 7	9	18	13	17	23	8	13	20
supplier 8	17	24	11	10	7	17	5	16

2,4								
	demand node 1	demand node 2	demand node 3	demand node 4	demand node 5	demand node 6	demand node 7	demand node 8
supplier 1	914	914	753	981	726	889	787	714
supplier 2	984	883	731	887	865	802	762	975
supplier 3	993	755	946	914	871	822	890	997
supplier 4	736	986	723	869	751	934	900	904
supplier 5	783	778	728	816	846	918	758	859
supplier 6	813	759	736	910	721	966	870	703
supplier 7	898	824	773	915	762	854	735	904
supplier 8	754	934	915	994	793	728	773	721
	demand node 1	demand node 2	demand node 3	demand node 4	demand node 5	demand node 6	demand node 7	demand node 8
supplier 1	52	53	46	35	47	53	47	63
supplier 2	25	51	70	74	46	29	43	26
supplier 3	28	40	44	56	78	54	67	43
supplier 4	58	46	57	50	57	62	64	56
supplier 5	72	50	69	57	62	46	46	48
supplier 6	79	24	74	69	63	71	62	23
supplier 7	54	74	76	52	41	64	77	33
supplier 8	80	23	31	32	51	41	67	70
	demand node 1	demand node 2	demand node 3	demand node 4	demand node 5	demand node 6	demand node 7	demand node 8
supplier 1	21	15	11	19	11	23	16	24
supplier 2	22	16	22	20	11	6	22	14
supplier 3	8	17	12	14	23	14	16	10
supplier 4	7	20	22	12	10	5	23	14
supplier 5	22	22	8	25	11	20	13	19
supplier 6	18	13	7	13	13	19	12	13
supplier 7	5	6	23	14	19	9	15	8
supplier 8	23	20	5	8	8	19	10	22

Table 6.2. *Transport costs, time and distance data*

The maximum capacity of the distribution centers and the maximum load of the means of transportation, where we have four types of transportation, are shown in Table 6.3.

L	8000	8000		
d	3900	3800	4200	4100

Table 6.3. *Data of distribution centers and means of transportation*

The data was obtained from the General Company for Trade in Iraqi Foodstuffs to determine transportation costs, time costs and distance costs for eight different demand centers, eight different display centers, two distribution lines and four means of transportation, as shown in Tables 6.1–6.3.

The aim of this study is to reduce transportation costs, time and distance. After applying the MATLAB program, we get the following results:

	K, q	demand node 1	demand node 2	demand node 3	demand node 4	demand node 5	demand node 6	demand node 7	demand node 8
supplier 1		0	0	0	0	0	0	0	0
supplier 2		0	0	0	0	0	0	0	0
supplier 3		0	0	0	0	0	0	0	0
supplier 4		0	0	0	0	0	0	0	0
supplier 5		0	0	0	0	0	0	0	0
supplier 6		0	0	0	0	0	0	0	0
supplier 7		100	0	0	0	0	0	0	0
supplier 8	(1,1)	0	0	0	0	0	0	0	0
	K, q	demand node 1	demand node 2	demand node 3	demand node 4	demand node 5	demand node 6	demand node 7	demand node 8
supplier 1		0	0	0	0	0	0	0	0
supplier 2		1500	0	0	0	0	0	0	0
supplier 3		0	0	0	0	0	0	0	0
supplier 4		0	0	0	0	0	0	0	0
supplier 5		0	0	0	0	0	0	0	0
supplier 6		0	0	300	0	0	0	0	0
supplier 7		0	0	0	0	0	0	0	0
supplier 8	(2,1)	0	0	0	2000	0	0	0	0

	K, q	demand node 1	demand node 2	demand node 3	demand node 4	demand node 5	demand node 6	demand node 7	demand node 8
supplier 1		0	0	0	0	0	0	0	0
supplier 2		0	0	0	0	0	0	0	0
supplier 3		0	0	0	0	0	0	0	0
supplier 4		0	0	0	0	0	0	0	0
supplier 5		0	0	0	0	0	0	0	0
supplier 6		0	0	0	0	0	0	0	0
supplier 7		0	0	0	0	0	0	0	0
supplier 8	(1,2)	0	0	0	0	0	0	0	0
	K, q	demand node 1	demand node 2	demand node 3	demand node 4	demand node 5	demand node 6	demand node 7	demand node 8
supplier 1		0	0	0	0	0	0	0	0
supplier 2		0	0	0	0	0	0	0	0
supplier 3		0	0	0	0	0	0	2000	0
supplier 4		0	0	0	0	0	0	0	0
supplier 5		0	0	0	0	0	0	100	0
supplier 6		0	0	0	0	0	0	0	0
supplier 7		0	0	0	0	0	0	0	1700
supplier 8	(2,2)	0	0	0	0	0	0	0	0
	K, q	demand node 1	demand node 2	demand node 3	demand node 4	demand node 5	demand node 6	demand node 7	demand node 8
supplier 1		0	300	0	0	0	0	0	0
supplier 2		0	0	0	0	0	0	0	0
supplier 3		0	0	0	0	0	0	0	0
supplier 4		0	0	0	0	2000	0	0	0
supplier 5		200	0	0	0	0	200	0	0
supplier 6		0	1300	0	0	0	0	0	0
supplier 7		0	0	0	0	200	0	0	0
supplier 8	(1,3)	0	0	0	0	0	0	0	0

	K, q	demand node 1	demand node 2	demand node 3	demand node 4	demand node 5	demand node 6	demand node 7	demand node 8
supplier 1		0	0	0	0	0	0	0	0
supplier 2		0	0	0	0	0	0	0	0
supplier 3		0	0	0	0	0	0	0	0
supplier 4		0	0	0	0	0	0	0	0
supplier 5		0	0	0	0	0	0	0	0
supplier 6		0	0	0	0	0	0	0	0
supplier 7		0	0	0	0	0	0	0	0
supplier 8	(2,3)	0	0	0	0	0	0	0	0
		demand node 1	demand node 2	demand node 3	demand node 4	demand node 5	demand node 6	demand node 7	demand node 8
supplier 1		0	0	1700	0	0	0	0	0
supplier 2		0	0	0	500	0	0	0	0
supplier 3		0	0	0	0	0	0	0	0
supplier 4		0	0	0	0	0	0	0	0
supplier 5		0	0	0	0	0	1500	0	0
supplier 6		0	0	0	0	0	0	0	0
supplier 7		0	0	0	0	0	0	0	0
supplier 8		0	0	0	0	0	0	0	0
		demand node 1	demand node 2	demand node 3	demand node 4	demand node 5	demand node 6	demand node 7	demand node 8
supplier 1		0	0	0	0	0	0	0	0
supplier 2		0	0	0	0	0	0	0	0
supplier 3		0	0	0	0	0	0	0	0
supplier 4		0	0	0	0	0	0	0	0
supplier 5		0	0	0	0	0	0	0	0
supplier 6		0	0	0	0	0	0	0	400
supplier 7		0	0	0	0	0	0	0	0
supplier 8		0	0	0	0	0	0	0	0

	K, q	demand node 1	demand node 2	demand node 3	demand node 4	demand node 5	demand node 6	demand node 7	demand node 8
supplier 1		0	0	1700	0	0	0	0	0
supplier 2		0	0	0	500	0	0	0	0
supplier 3		0	0	0	0	0	0	0	0
supplier 4		0	0	0	0	0	0	0	0
supplier 5		0	0	0	0	0	1500	0	0
supplier 6		0	0	0	0	0	0	0	0
supplier 7		0	0	0	0	0	0	0	0
supplier 8	(1,4)	0	0	0	0	0	0	0	0
	K, q	demand node 1	demand node 2	demand node 3	demand node 4	demand node 5	demand node 6	demand node 7	demand node 8
supplier 1		0	0	0	0	0	0	0	0
supplier 2		0	0	0	0	0	0	0	0
supplier 3		0	0	0	0	0	0	0	0
supplier 4		0	0	0	0	0	0	0	0
supplier 5		0	0	0	0	0	0	0	0
supplier 6		0	0	0	0	0	0	0	400
supplier 7		0	0	0	0	0	0	0	0
supplier 8	(2,4)	0	0	0	0	0	0	0	0

Table 6.4. *Results of the amounts of material transferred for SPEA2*

After substituting into equations (1-1) (1-2) (1-3), the least costs, the least times and the least distance are obtained, i.e. by multiplying the matrices (1,1) (2,1) (1,2) (2,2) (1,3) (2,3) (1,4) (2,4) and costs of transporting one unit $C_{i,j,k,q}$ from the materials corresponding to the matrices (1,1) (2,1) (1,2) (2,2) (1,3) (2,3) (1,4) (2,4) regarding the amount of different transported materials $X_{i,j,k,q}$:

cost function= 8,839,000;

time function= 507,800;

distance function= 244,600.

Subject to:

As for the restrictions related to the quantities of materials available in the supply centers and according to equation (1-4), we have eight restrictions, and as shown below after collecting each of the eight different supply centers for all the quantities of different transported materials.

(1,1) (2,1)

supplier 1	0	0	0	0	0	0	0	0	+	supplier 1	0	0	0	0	0	0	0	0

(1,2) (2,2)

supplier 1	0	0	0	0	0	0	0	0	+	supplier 1	0	0	0	0	0	0	0	0

(1,3) (2,3)

supplier 1	0	300	0	0	0	0	0	0	+	supplier 1	0	0	0	0	0	0	0	0

(1,4) (2,4)

supplier 1	0	0	1700	0	0	0	0	0	+	supplier 1	0	0	0	0	0	0	0	0

This means that the total quantity transported from supplier 1 is equal to the available quantity (2,000).

(1,1) (2,1)

supplier 2	0	0	0	0	0	0	0	0	+	supplier 2	1500	0	0	0	0	0	0	0

(1,2) (2,2)

supplier 2	0	0	0	0	0	0	0	0	+	supplier 2	0	0	0	0	0	0	0	0

(1,3) (2,3)

supplier 2	0	0	0	0	0	0	0	0	+	supplier 2	0	0	0	0	0	0	0	0

(1,4) (2,4)

supplier 2	0	0	0	500	0	0	0	0	+	supplier 2	0	0	0	0	0	0	0	0

This means that the total quantity transported from the supplier 2 is equal to the available quantity (2,000).

(1,1) (2,1)

supplier 3	0	0	0	0	0	0	0	0	+	supplier 3	0	0	0	0	0	0	0	0

(1,2) (2,2)

supplier 3	0	0	0	0	0	0	0	0	+	supplier 3	0	0	0	0	0	0	2000	0

(1,3) (2,3)

supplier 3	0	0	0	0	0	0	0	0	+	supplier 3	0	0	0	0	0	0	0	0

(1,4) (2,4)

supplier 3	0	0	0	0	0	0	0	0	+	supplier 3	0	0	0	0	0	0	0	0

This means that the total quantity transported from supplier 3 is equal to the available quantity (2,000).

				(1,1)										(2,1)				
supplier 4	0	0	0	0	0	0	0	0	+	**supplier 4**	0	0	0	0	0	0	0	0
				(1,2)										(2,2)				
supplier 4	0	0	0	0	0	0	0	0	+	**supplier 4**	0	0	0	0	0	0	0	0
				(1,3)										(2,3)				
supplier 4	0	0	0	0	2000	0	0	0	+	**supplier 4**	0	0	0	0	0	0	0	0
				(1,4)										(2,4)				
supplier 4	0	0	0	0	0	0	0	0	+	**supplier 4**	0	0	0	0	0	0	0	0

This means that the total quantity transported from supplier 4 is equal to the available quantity (2,000).

				(1,1)										(2,1)				
supplier 5	0	0	0	0	0	0	0	0	+	**supplier 5**	0	0	0	0	0	0	0	0
				(1,2)										(2,2)				
supplier 5	0	0	0	0	0	0	0	0	+	**supplier 5**	0	0	0	0	0	0	100	0
				(1,3)										(2,3)				
supplier 5	200	0	0	0	0	200	0	0	+	**supplier 5**	0	0	0	0	0	0	0	0
				(1,4)										(2,4)				
supplier 5	0	0	0	0	0	1500	0	0	+	**supplier 5**	0	0	0	0	0	0	0	0

This means that the total quantity transported from supplier 5 is equal to the available quantity (2,000).

				(1,1)										(2,1)				
supplier 6	0	0	0	0	0	0	0	0	+	**supplier 6**	0	0	300	0	0	0	0	0
				(1,2)										(2,2)				
supplier 6	0	0	0	0	0	0	0	0	+	**supplier 6**	0	0	0	0	0	0	0	0
				(1,3)										(2,3)				
supplier 6	0	1300	0	0	0	0	0	0	+	**supplier 6**	0	0	0	0	0	0	0	0
				(1,4)										(2,4)				
supplier 6	0	0	0	0	0	0	0	0	+	**supplier 6**	0	0	0	0	0	0	0	400

This means that the total quantity transported from supplier 6 is equal to the available quantity (2,000).

	(1,1)										(2,1)							
supplier 7	100	0	0	0	0	0	0	0	+	**supplier 7**	0	0	0	0	0	0	0	0
	(1,2)										(2,2)							
supplier 7	0	0	0	0	0	0	0	0	+	**supplier 7**	0	0	0	0	0	0	0	1700
	(1,3)										(2,3)							
supplier 7	0	0	0	0	200	0	0	0	+	**supplier 7**	0	0	0	0	0	0	0	0
	(1,4)										(2,4)							
supplier 7	0	0	0	0	0	0	0	0	+	**supplier 7**	0	0	0	0	0	0	0	0

This means that the total quantity transported from supplier 7 is equal to the available quantity (2,000).

	(1,1)										(2,1)							
supplier 8	0	0	0	0	0	0	0	0	+	**supplier 8**	0	0	0	2000	0	0	0	0
	(1,2)										(2,2)							
supplier 8	0	0	0	0	0	0	0	0	+	**supplier 8**	0	0	0	0	0	0	0	0
	(1,3)										(2,3)							
supplier 8	0	0	0	0	0	0	0	0	+	**supplier 8**	0	0	0	0	0	0	0	0
	(1,4)										(2,4)							
supplier 8	0	0	0	0	0	0	0	0	+	**supplier 8**	0	0	0	0	0	0	0	0

This means that the total quantity transported from supplier 8 is equal to the available quantity (2,000).

Also, for the restrictions of the different demand centers, we have eight restrictions. According to equation (1-5), the eight different demand centers were collected for the different quantities of transported materials, as shown below:

(1,1)		(2,1)		(1,2)		(2,2)		(1,3)		(2,3)		(1,4)		(2,4)
demand node 1		demand node 1		demand node 1		demand node 1		demand node 1		demand node 1		demand node 1		demand node 1
0		0		0		0		0		0		0		0
0		1500		0		0		0		0		0		0
0	+	0	+	0	+	0	+	0	+	0	+	0	+	0
0		0		0		0		0		0		0		0
0		0		0		0		200		0		0		0
0		0		0		0		0		0		0		0
100		0		0		0		0		0		0		0
0		0		0		0		0		0		0		0

This means that the transported quantity for the first-order center is equal to its order and it does not need a transfer from another center.

(1,1)		(2,1)		(1,2)		(2,2)		(1,3)		(2,3)		(1,4)		(2,4)
demand node 2		demand node 2		demand node 2		demand node 2		demand node 2		demand node 2		demand node 2		demand node 2
0		0		0		0		300		0		0		0
0		0		0		0		0		0		0		0
0	+	0	+	0	+	0	+	0	+	0	+	0	+	0
0		0		0		0		0		0		0		0
0		0		0		0		0		0		0		0
0		0		0		0		1300		0		0		0
0		0		0		0		0		0		0		0
0		0		0		0		0		0		0		0

This means that the transported quantity for the second-order center is equal to its order and it does not need a transfer from another center.

(1,1)		(2,1)		(1,2)		(2,2)		(1,3)		(2,3)		(1,4)		(2,4)
demand node 3		demand node 3		demand node 3		demand node 3		demand node 3		demand node 3		demand node 3		demand node 3
0		0		0		0		0		0		1700		0
0		0		0		0		0		0		0		0
0	+	0	+	0	+	0	+	0	+	0	+	0	+	0
0		0		0		0		0		0		0		0
0		0		0		0		0		0		0		0
0		300		0		0		0		0		0		0
0		0		0		0		0		0		0		0
0		0		0		0		0		0		0		0

This means that the transported quantity for the third-order center is equal to its order and it does not need a transfer from another center.

(1,1)		(2,1)		(1,2)		(2,2)		(1,3)		(2,3)		(1,4)		(2,4)
demand node 4		demand node 4		demand node 4		demand node 4		demand node 4		demand node 4		demand node 4		demand node 4
0		0		0		0		0		0		0		0
0		0		0		0		0		0		500		0
0	+	0	+	0	+	0	+	0	+	0	+	0	+	0
0		0		0		0		0		0		0		0
0		0		0		0		0		0		0		0
0		0		0		0		0		0		0		0
0		0		0		0		0		0		0		0
0		2000		0		0		0		0		0		0

This means that the transported quantity for the fourth-order center is equal to its order and it does not need a transfer from another center.

(1,1)		(2,1)		(1,2)		(2,2)		(1,3)		(2,3)		(1,4)		(2,4)
demand node 5		demand node 5		demand node 5		demand node 5		demand node 5		demand node 5		demand node 5		demand node 5
0		0		0		0		0		0		0		0
0		0		0		0		0		0		0		0
0	+	0	+	0	+	0	+	0	+	0	+	0	+	0
0		0		0		0		2000		0		0		0
0		0		0		0		0		0		0		0
0		0		0		0		0		0		0		0
0		0		0		0		200		0		0		0
0		0		0		0		0		0		0		0

This means that the transported quantity for the fifth-order center is equal to its order and it does not need a transfer from another center.

(1,1)		(2,1)		(1,2)		(2,2)		(1,3)		(2,3)		(1,4)		(2,4)
demand node 6		demand node 6		demand node 6		demand node 6		demand node 6		demand node 6		demand node 6		demand node 6
0		0		0		0		0		0		0		0
0		0		0		0		0		0		0		0
0	+	0	+	0	+	0	+	0	+	0	+	0	+	0
0		0		0		0		0		0		0		0
0		0		0		0		200		0		1500		0
0		0		0		0		0		0		0		0
0		0		0		0		0		0		0		0
0		0		0		0		0		0		0		0

This means that the transported quantity for the sixth-order center is equal to his order and it does not need a transfer from another center.

(1,1)		(2,1)		(1,2)		(2,2)		(1,3)		(2,3)		(1,4)		(2,4)
demand node 7		demand node 7		demand node 7		demand node 7		demand node 7		demand node 7		demand node 7		demand node 7
0		0		0		0		0		0		0		0
0		0		0		0		0		0		0		0
0	+	0	+	0	+	2000	+	0	+	0	+	0	+	0
0		0		0		0		0		0		0		0
0		0		0		100		0		0		0		0
0		0		0		0		0		0		0		0
0		0		0		0		0		0		0		0
0		0		0		0		0		0		0		0

This means that the transported quantity for the seventh-order center is equal to his order and it does not need a transfer from another center.

(1,1)		(2,1)		(1,2)		(2,2)		(1,3)		(2,3)		(1,4)		(2,4)
demand node 8		demand node 8		demand node 8		demand node 8		demand node 8		demand node 8		demand node 8		demand node 8
0		0		0		0		0		0		0		0
0		0		0		0		0		0		0		0
0	+	0	+	0	+	0	+	0	+	0	+	0	+	0
0		0		0		0		0		0		0		0
0		0		0		0		0		0		0		0
0		0		0		0		0		0		0		400
0		0		0		1700		0		0		0		0
0		0		0		0		0		0		0		0

This means that the transported quantity for the eighth-order center is equal to his order and it does not need a transfer from another center.

For the constraints of the distribution centers, we have two constraints, according to equation [6.1] to [6.7].

It should be ensured that each distribution center has the quantities entering it equal to 8,000. In this case, the first entry results from the matrix summation of the quantities of transported materials (1,1) (1,2) (1,3) (1,4).

	Matrix of the amount of material transported (1,1)									Matrix of the amount of material transported (1,2)							
	demand node 1	demand node 2	demand node 3	demand node 4	demand node 5	demand node 6	demand node 7	demand node 8		demand node 1	demand node 2	demand node 3	demand node 4	demand node 5	demand node 6	demand node 7	demand node 8
supplier 1	0	0	0	0	0	0	0	0		0	0	0	0	0	0	0	0
supplier 2	0	0	0	0	0	0	0	0		0	0	0	0	0	0	0	0
supplier 3	0	0	0	0	0	0	0	0		0	0	0	0	0	0	0	0
supplier 4	0	0	0	0	0	0	0	0	+	0	0	0	0	0	0	0	0
supplier 5	0	0	0	0	0	0	0	0		0	0	0	0	0	0	0	0
supplier 6	0	0	0	0	0	0	0	0		0	0	0	0	0	0	0	0
supplier 7	100	0	0	0	0	0	0	0		0	0	0	0	0	0	0	0
supplier 8	0	0	0	0	0	0	0	0		0	0	0	0	0	0	0	0

	Matrix of the amount of material transported(1,3)									Matrix of the amount of material transported(1,4)							
	demand node 1	demand node 2	demand node 3	demand node 4	demand node 5	demand node 6	demand node 7	demand node 8		demand node 1	demand node 2	demand node 3	demand node 4	demand node 5	demand node 6	demand node 7	demand node 8
supplier 1	0	300	0	0	0	0	0	0		0	0	1700	0	0	0	0	0
supplier 2	0	0	0	0	0	0	0	0		0	0	0	500	0	0	0	0
supplier 3	0	0	0	0	0	0	0	0		0	0	0	0	0	0	0	0
supplier 4	0	0	0	0	2000	0	0	0	+	0	0	0	0	0	0	0	0
supplier 5	200	0	0	0	0	200	0	0		0	0	0	0	0	1500	0	0
supplier 6	0	1300	0	0	0	0	0	0		0	0	0	0	0	0	0	0
supplier 7	0	0	0	0	200	0	0	0		0	0	0	0	0	0	0	0
supplier 8	0	0	0	0	0	0	0	0		0	0	0	0	0	0	0	0

For the restrictions related to the means of transportation, we have four restrictions, according to equation [6.6].

	Matrix of the amount of material transported(2,1)									Matrix of the amount of material transported(2,2)							
	demand node 1	demand node 2	demand node 3	demand node 4	demand node 5	demand node 6	demand node 7	demand node 8		demand node 1	demand node 2	demand node 3	demand node 4	demand node 5	demand node 6	demand node 7	demand node 8
supplier 1	0	0	0	0	0	0	0	0		0	0	0	0	0	0	0	0
supplier 2	1500	0	0	0	0	0	0	0		0	0	0	0	0	0	0	0
supplier 3	0	0	0	0	0	0	0	0		0	0	0	0	0	0	2000	0
supplier 4	0	0	0	0	0	0	0	0	+	0	0	0	0	0	0	0	0
supplier 5	0	0	0	0	0	0	0	0		0	0	0	0	0	0	100	0
supplier 6	0	0	300	0	0	0	0	0		0	0	0	0	0	0	0	0
supplier 7	0	0	0	0	0	0	0	0		0	0	0	0	0	0	0	1700
supplier 8	0	0	0	2000	0	0	0	0		0	0	0	0	0	0	0	0

	Matrix of the amount of material transported(2,3)									Matrix of the amount of material transported(2,4)							
	demand node 1	demand node 2	demand node 3	demand node 4	demand node 5	demand node 6	demand node 7	demand node 8		demand node 1	demand node 2	demand node 3	demand node 4	demand node 5	demand node 6	demand node 7	demand node 8
supplier 1	0	0	0	0	0	0	0	0		0	0	0	0	0	0	0	0
supplier 2	0	0	0	0	0	0	0	0		0	0	0	0	0	0	0	0
supplier 3	0	0	0	0	0	0	0	0		0	0	0	0	0	0	0	0
supplier 4	0	0	0	0	0	0	0	0	+	0	0	0	0	0	0	0	0
supplier 5	0	0	0	0	0	0	0	0		0	0	0	0	0	0	0	0
supplier 6	0	0	0	0	0	0	0	0		0	0	0	0	0	0	0	400
supplier 7	0	0	0	0	0	0	0	0		0	0	0	0	0	0	0	0
supplier 8	0	0	0	0	0	0	0	0		0	0	0	0	0	0	0	0

For the first constraint, we collect each of the matrices for the quantities of transported materials (1,1) (2,1) to find out the capacity of the first means of transport, as shown below:

	Matrix of the amount of material transported(1,1)									Matrix of the amount of material transported(2,1)							
	demand node 1	demand node 2	demand node 3	demand node 4	demand node 5	demand node 6	demand node 7	demand node 8		demand node 1	demand node 2	demand node 3	demand node 4	demand node 5	demand node 6	demand node 7	demand node 8
supplier 1	0	0	0	0	0	0	0	0		0	0	0	0	0	0	0	0
supplier 2	0	0	0	0	0	0	0	0		1500	0	0	0	0	0	0	0
supplier 3	0	0	0	0	0	0	0	0		0	0	0	0	0	0	0	0
supplier 4	0	0	0	0	0	0	0	0	+	0	0	0	0	0	0	0	0
supplier 5	0	0	0	0	0	0	0	0		0	0	0	0	0	0	0	0
supplier 6	0	0	0	0	0	0	0	0		0	0	300	0	0	0	0	0
supplier 7	100	0	0	0	0	0	0	0		0	0	0	0	0	0	0	0
supplier 8	0	0	0	0	0	0	0	0		0	0	0	2000	0	0	0	0

For the second constraint, we collect each of the matrices for the quantities of transported materials (1,2) (2,2) to find out the capacity of the second means of transport, as shown below:

	Matrix of the amount of material transported(1,2)									Matrix of the amount of material transported(2,2)							
	demand node 1	demand node 2	demand node 3	demand node 4	demand node 5	demand node 6	demand node 7	demand node 8		demand node 1	demand node 2	demand node 3	demand node 4	demand node 5	demand node 6	demand node 7	demand node 8
supplier 1	0	0	0	0	0	0	0	0		0	0	0	0	0	0	0	0
supplier 2	0	0	0	0	0	0	0	0		0	0	0	0	0	0	0	0
supplier 3	0	0	0	0	0	0	0	0		0	0	0	0	0	0	2000	0
supplier 4	0	0	0	0	0	0	0	0	+	0	0	0	0	0	0	0	0
supplier 5	0	0	0	0	0	0	0	0		0	0	0	0	0	0	100	0
supplier 6	0	0	0	0	0	0	0	0		0	0	0	0	0	0	0	0
supplier 7	0	0	0	0	0	0	0	0		0	0	0	0	0	0	0	1700
supplier 8	0	0	0	0	0	0	0	0		0	0	0	0	0	0	0	0

For the second constraint, we obtain the result of the matrix summation of the quantities of transported materials (2,1) (2,2) (2,3) (2,4).

	Matrix of the amount of material transported(1,3)									Matrix of the amount of material transported(2,3)							
	demand node 1	demand node 2	demand node 3	demand node 4	demand node 5	demand node 6	demand node 7	demand node 8		demand node 1	demand node 2	demand node 3	demand node 4	demand node 5	demand node 6	demand node 7	demand node 8
supplier 1	0	300	0	0	0	0	0	0		0	0	0	0	0	0	0	0
supplier 2	0	0	0	0	0	0	0	0		0	0	0	0	0	0	0	0
supplier 3	0	0	0	0	0	0	0	0		0	0	0	0	0	0	0	0
supplier 4	0	0	0	0	2000	0	0	0	+	0	0	0	0	0	0	0	0
supplier 5	200	0	0	0	0	200	0	0		0	0	0	0	0	0	0	0
supplier 6	0	1300	0	0	0	0	0	0		0	0	0	0	0	0	0	0
supplier 7	0	0	0	0	200	0	0	0		0	0	0	0	0	0	0	0
supplier 8	0	0	0	0	0	0	0	0		0	0	0	0	0	0	0	0

For the third constraint, we collect each of the matrices of the quantities of transported materials (1,3) (2,3) to find out the capacity of the third means of transport, as shown below:

	Matrix of the amount of material transported(1,4)									Matrix of the amount of material transported(2,4)							
	demand node 1	demand node 2	demand node 3	demand node 4	demand node 5	demand node 6	demand node 7	demand node 8		demand node 1	demand node 2	demand node 3	demand node 4	demand node 5	demand node 6	demand node 7	demand node 8
supplier 1	0	0	1700	0	0	0	0	0		0	0	0	0	0	0	0	0
supplier 2	0	0	0	500	0	0	0	0		0	0	0	0	0	0	0	0
supplier 3	0	0	0	0	0	0	0	0		0	0	0	0	0	0	0	0
supplier 4	0	0	0	0	0	0	0	0	+	0	0	0	0	0	0	0	0
supplier 5	0	0	0	0	0	1500	0	0		0	0	0	0	0	0	0	0
supplier 6	0	0	0	0	0	0	0	0		0	0	0	0	0	0	0	400
supplier 7	0	0	0	0	0	0	0	0		0	0	0	0	0	0	0	0
supplier 8	0	0	0	0	0	0	0	0		0	0	0	0	0	0	0	0

For the fourth constraint, we collect each of the matrices for the quantities of transported materials (1,4) (2.4) to find out the capacity of the fourth means of transport, as shown below:

Best solution using the SPEA2:

decoding	3	19	8	7	5	6	13	15	9	11	2	20	4	10	22	21	12	16	17	18	1	14

6.5. Results

The results were obtained using the MATLAB 2020 program.

By solving the 4D transportation model, the lowest cost (8,839,000 dollars), the least time (507,800 hours) and the least distance (244,600 km) were obtained through the different distribution lines for the different means of transportation from the different supply centers to the different demand centers.

6.6. Conclusions and perspectives

In this chapter, we formulated the 4D transportation problem as a multi-objective linear program and solved it by the metaheuristic SPEA2.

Different results were obtained using the proposed algorithm of the multi-objective 4D transportation model for the data obtained from the General Company for Food Stuff Trading by taking different samples (eight supply centers, eight demand centers, two distribution lines, four different means of transportation). The results were obtained using the MATLAB program, and we chose the best solution among those multiple optimal solutions by means of the decoding procedure.

The results showed that extracting the uncontrolled points from multiple runs of the algorithm provides better results than a single run.

We recommend the following points:

1) The generalization of the multi-objective 4D transportation model to all companies that require decisions to reduce transport costs, time and distance, whether the different means of transport through the different distribution lines in addition to the different supply and demand centers.

2) The developing of the 4D transportation model to include other variables such as the cost of customer satisfaction, stock level (maintaining the lowest level of stock), expansion by opening other branches.

3) The use of other objectives to the transfer model such as profits, customer satisfaction level, stock ratio, expansion by opening other branches and using other algorithms to solve.

4) Linking the issue of transportation with rotation or transportation in several stages.

5) Expand research and develop its own algorithms by supporting and encouraging researchers in this direction, as these methods have proven to achieve the goals of the model.

6) Giving special courses to employees and decision-makers in the Ministry of Commerce on the MATLAB program.

Future work will be to solve the multi-objective 5D transportation model and generalize its use to all companies that require decisions to reduce transportation costs, times and distance together, whether the different transportation means are used across the different distribution lines in addition to homogeneous or heterogeneous materials. We also propose to compare several different metaheuristics for solving the metaheuristics.

6.7. References

Anderson, E.W., Fornell, C., Rust, R.T. (1997). Customer satisfaction, productivity, and profitability: Differences between goods and services. *Marketing Science*, 16(2), 129–145.

Bakhayt, A.G.K. (2016). Problem solving bi-objective 4-dimensional transportation by using PSO. *Science International*, 28(3), 2403–2410.

Bera, S., Giri, P.K., Jana, D.K., Basu, K., Maiti, M. (2018). Multi-item 4D-TPs under budget constraint using rough interval. *Applied Soft Computing*, 71, 364–385.

Costa, J.P.A., Cortes, O.A.C., Júnior, E.C., Vargas, A.G., Castelo, M. (2017). An adaptive algorithm for updating populations on (SPEA2). *Simpósio Brasileiro de Automaçao Inteligente (SBAI)*.

Halder, J.S., Das, B., Panigrahi, G., Maiti, M. (2017). Some special fixed charge solid transportation problems of substitute and breakable items in crisp and fuzzy environments. *Computers and Industrial Engineering*, 111, 272–281.

Halder J.S., Das, B., Panigrahi, G., Maiti, M. (2019). Constrained FC 4D MITPs for damageable substitutable and complementary items in rough environments. *Mathematics*, 7(3), 281.

Haley, K.B. (1962). New methods in mathematical programming–The solid transportation problem. *Operations Research*, 10(4), 448–463.

Kim, M., Hiroyasu, T., Miki, M., Watanabe, S. (2004). SPEA2+: Improving the performance of the strength Pareto evolutionary algorithm 2. *International Conference on Parallel Problem Solving from Nature*, 742–751. Springer, Berlin.

Mukherjee, S., Pal, J., Bagchi, S. (2021). On decoding procedures of intertwining codes. *Journal of Discrete Mathematical Sciences and Cryptography*, 24(4), 955–964.

Sahoo, P., Jana, D.K., Pramanik, S., Panigrahi, G. (2020). Uncertain four-dimensional multi-objective multi-item transportation models via GP technique. *Soft Computing*, 24(22), 17291–17307.

Zitzler, E. and Thiele, L. (1998). An evolutionary algorithm for multiobjective optimization: The strength Pareto approach. TIK-Report, 43.

Zitzler, E., Laumanns, M., Thiele, L. (2001). SPEA2: Improving the strength Pareto evolutionary algorithm. TIK-Report, 103.

7

Comparison of Classical and Machine Learning Forecasting Techniques: A Case Study of Rice Production in India

The ability to forecast well-founded market information, and notable production predictions for agricultural commodities, is critical for the farming community and the government's long-term success. This type of prediction is generally performed on the time series data by using different forecasting models. This study uses classical methods such as SMA (Simple Moving Average), WMA (Weighted Moving Average), ES (Exponential Smoothing), ARIMA (Autoregressive Integrated Moving Average) and machine learning techniques such as LR (Linear Regression), ANN (Artificial Neural Network), SVR (Support Vector Regressor), RF (Random Forest), KNN (k-nearest neighbors) and FB-Prophet for forecasting agriculture production based on time series data. The rice production time-series data were taken for a period of 73 years (1949–2021) in India. The methods used for forecasting in this study are compared through standard performance metrics such as MAPE (Mean Absolute Percentage Error), RMSE (Root Mean Square Error) and MAE (Mean Absolute Error). It was observed that RF performed better in comparison to other models and forecasted the value for rice production to be 108,892.7 thousand tons for 2022.

7.1. Introduction

India is the land of agriculture and ranks second in the world regarding agricultural products. About 66.67% of the total population is dependent on agriculture directly and indirectly (Das et al. 2009). According to the "Agriculture

Chapter written by Deepanshu NAYAK, Millie PANT and Kailash Chandra LACHHWANI.

Census of India" survey in 2011, agriculture is the source of income for 62% of the Indian population living in rural areas. In addition, agriculture's contribution to India's GDP is about 18% (Hellerstein and Scott 2011).

For the efficient distribution of agri-food production (AFP) and in order to benefit the farmers in terms of profit, it is essential to develop an efficient supply chain of the AFP. A supply chain is a flow of material from the producer to the consumer with different stages in between. In the agri-food supply chain (AFSC), farmers may be treated as producers, producing crops in their fields (Yadav et al. 2022).

An AFSC consists of the following operations:

1) Manufacturing/producing: production or manufacture of the final product.

2) Forecasting: prediction of demand and production with respect to the consumer and the producer.

3) Transporting: flow of material from within and outside the production.

4) Managing the inventory: storing the materials and products at different levels.

5) Market: supplying the product in the market.

6) Consumer: final destination of the product.

Forecasting plays a crucial role for policymakers, farmers and governmental offices as most other operations of AFSC depend directly or indirectly on it. While efficient forecasting will help the farmers in deciding the type and quantity of crops to be grown, it will also help government agencies in making decisions in favor of the farmers, etc.

The different stages of AFSC are shown in Figure 7.1.

According to the FAO (Food and Agriculture Organization of the United Nations), India is the second-largest producer of rice in the world[1]. Uttar-Pradesh and West Bengal are the two states of India which are the highest contributors (14% each) in the production of rice (Reserve Bank of India 2021) in India. An efficient prediction of the production of rice can help us to estimate the consumption quantity, storage-space requirements and the quantity that can be exported. The three major areas that will benefit from the forecasting of rice production are shown in Figure 7.3.

1 FAO (2022). FAO in India: India at a glance. Food and Agriculture Organization of the United Nations [Online]. Available at: https://www.fao.org/india/fao-in-india/india-at-a-glance/en/#:~:text=India%20is%20the%20world's%20largest,poultry%2C%20livestock%20and%20plantation%20crops.

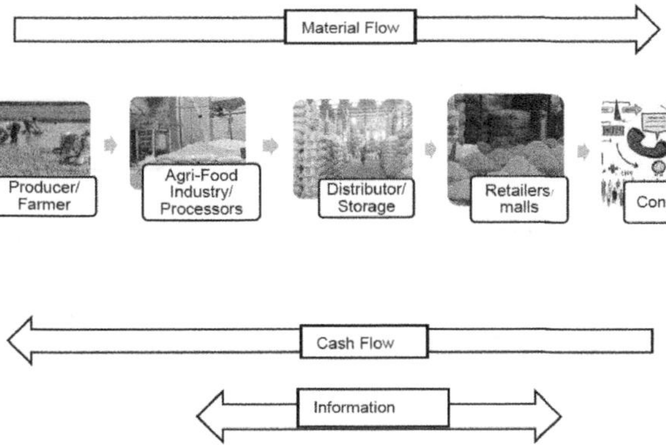

Figure 7.1. *Agri-food supply chain structure[2,3,4,5,6]. For a color version of this figure, see www.iste.co.uk/koubaa/optimization.zip*

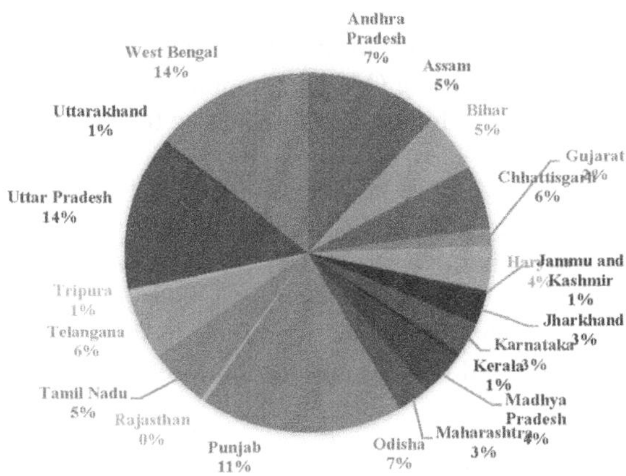

Figure 7.2. *State-wise production of rice in India (see Footnote 1). For a color version of this figure, see www.iste.co.uk/koubaa/optimization.zip*

2 Rice | Rice Exporter | Rice Distributor | Rice Wholesaler | Rice Supplier | Rice Importer | Basmati Rice | Rice For Sale | Long Grain Rice Exporter | Buy Rice Online | Rice For Sale | Basmati Rice Exporter | Basmati Rice Wholesaler | Long Grain Rice buyer. By Trans Merchant Trading Co Ltd., Malaysia.

3 India's rice production could hit record high as farmers expand their areas.

4 The problems we may encounter when operating rice milling machinery.

5 Benjamin, J. (2015). Three steps to better understanding your customer base. AllBusiness.com [Online]. Available at: https://www.allbusiness.com/three-steps-better-understanding-customer-base-24079-1.html.

6 Import Export Blog (2017). The buyer's guide to buying rice in bulk from suppliers around the world. Fomexco [Online]. Avaiable at: https://fomexco.com/blog-xuat-nhap-khau/the-buyers-guide-to-buying-rice-in-bulk-from-suppliers-around-the-world-138.

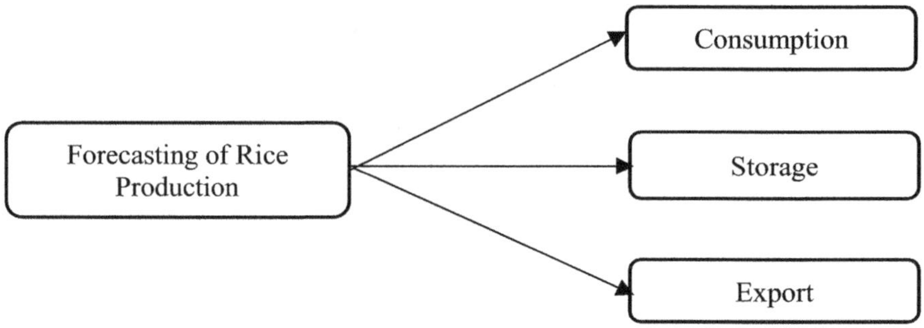

Figure 7.3. *Major areas that will benefit from forecasting*

Many research studies have been proposed for the forecasting of prices. Also, many classical and machine learning (ML) techniques are used to make predictions. However, only a few studies are focused on forecasting the production.

In the present study, 10 methods are used for forecasting the production. These include: the simple moving average (SMA) method; weighted average (WA) method; exponential smoothing average (ESA); ARIMA (Autoregressive Integrated Moving Average); and ML techniques such as linear regression (LR), artificial neural network (ANN), support vector regressor (SVR), random forest (RF), k-nearest neighbors (KNN) and FB-Prophet for the forecasting of rice production in India.

Comparisons between methods are made using standard performance metrics such as mean absolute percentage error (MAPE), root mean square error (RMSE) and mean absolute error (MAE).

In section 7.2, a literature review on forecasting in agriculture is presented. Section 7.3 describes the methods and material used. Section 7.4 presents the results and discussions of the present study. In the final section 7.5, conclusions are made and future scopes are discussed.

References	Commodity	Forecasting techniques	Data source	Language used	Data size	Performance metrics
Mahto et al. (2021)	Sunflower seed and soybean seed	ANN, ARIMA	data.gov.in	Python	Monthly data of five years (January 2011–December 2016)	MAPE using ANN: Soybean = 2.4% Sunflower = 7.7% MAPE using ARIMA: Soybean = 19.76% Sunflower = 15.2% RMSPE using ANN: Soybean = 3.15% Sunflower = 8.92% RMSPE using ARIMA: Soybean = 19.84% Sunflower = 15.9%
Banerjee et al. (2022)	Cabbage, cauliflower and brinjal	ARIMA, LSTM	https://nhb.gov.in, Federal Reserve Bank of St. Louis website, https://mausam.imd.gov.in	Keras in Python	Monthly data of 15 years (2005–2020)	Best value of MAPE using ARIMA: 0.17 best value of RMSE using ARIMA: 3.21 Best value of MAPE using LSTM: 0.17 best value of RMSE using LSTM: 3.24
Sabu and Kumar (2020)	Areca nuts	SARIMA, Holt-Winter's seasonal method and LSTM neural network	Department of Economics and Statistics, Kerala	Python	Monthly data of 10 years (2007–2017)	RMSE values: SARIMA – 16.5475 Holt-Winter's seasonal method – 18.0589 LSTM (stationary data) – 7.2780
Li et al. (2021)	Vegetable	Optimized neural network combined model based on the IWOA operator	Beijing Agricultural and Rural Bureau	MATLAB (2018a, Mathworks, Inc.)	Monthly data of 10 years (January 2010–April 2020)	MAE = 0.175, RMSE = 0.226, MAPE = 0.062 TIC = 0.039
Ribeiro and dos Santos Coelho (2020)	Soybean and wheat	GBM, XGB, RF and STACK regression ensembles, as well as the Multilayer Perception Neural Network (MLP) SVR and KNN models	https://www.quandl.com	R Language	Monthly data of 17 years (July 2001–May 2018)	MAPE < 1% XGB/STACK (Least Absolute Shrinkage and Selection Operator – KNN–XGB–SVR) and RF models showed better performance for short-term forecasting

Zhang et al. (2020)	Piglet, hog, beef	ANN, SVR, ELM (Extreme Learning Machine), SMA, MSH-RF, MSN-RF, MSH-SVM, MSN-SVM MSH-FR-RF	http://www.caaa.cn/market	Python	522 monthly agricultural commodity price series	MAPE for all of the techniques used: ANN: 8.67, SVR: 10.45, ELM: 9.42, SMA: 8.68, MSN-RF: 8.66, MSN-SVM: 8.75, MSH-RF: 8.35, MSH-SVM: 8.49, MSH-FR-RF: 8.33
Anggraeni et al. (2017)	Rice	ARIMAX and VAR models	–	–	Monthly period data from 2000 to 2015	ARIMAX model predicts the rice consumer price with MAPE 0.15%, which is 15.27% better than the VAR model
Xu and Zhang (2021)	Corn	Neural network	GeoGrain Inc.	MATLAB	July 19, 2006–February 17, 2011	The mean RMSE is 0.00061 for training and 0.00057 for testing
Weng et al. (2019)	Horticultural products	RNN, ARIMA	Beijing Xinfadi	Python	Monthly, weekly and daily (August 19, 2015 to July 30, 2018)	Average absolute error ARIMA = 0.55 CNY RNN = 0.49 CNY

Table 7.1. *Studies related to the forecasting of agri-food products*

7.2. Literature review

In Mahto et al. (2021), the authors proposed a study on time-series data for the price of soybean and sunflower by using ANN and ARIMA models. On computation, MAPE and root mean square percentage error (RMSPE) for soybean using the ANN is reported to be 2.4% and 3.4%, respectively, and by using ARIMA as 3.15% and 19.76%, respectively. Similarly, for the sunflower, these values are 7.7%, 15.2%, 8.92% and 15.9%, respectively. In Banerjee et al. (2022), the authors applied LR, KNN, decision tree, ANN, RF, SVR, ARIMA, Univariate LSTM (Long Short-Term Memory), Markov Switching Multivariate LSTM for the prediction of the price of Cabbage, Cauliflower and Brinjal. The authors have taken six cities in India and calculated that the minimum values of MSE (Mean Square Error), and RMSE are 1.52%, 2.32% and 0.13%, respectively, by using the proposed approaches. In Sabu and Kumar (2020), the authors predicted the monthly prices of areca nut in the Kerala state of India by using the SARIMA (Seasonal Autoregressive Integrated Moving Average), Holt-Winter's seasonal method, LSTM neural network, and reported the RMSE values as 16.547%, 18.0589% and 7.2780, respectively. For the vegetable price prediction in Li et al. (2021), the authors proposed a model for which MAE, RMSE, MAPE and TIC (Theil inequality coefficient) are 0.175, 0.226, 0.062 and 0.039, respectively. In Ribeiro and dos Santos Coelho (2020), the authors proposed Ensemble method based on bagging, boosting and stacking to predict the prices of soybean and wheat. This methodology presented a good performance with high accuracy and small proportional errors (MAPE < 1%) for both case studies. In Zhang et al. (2020), the authors used ML techniques and concluded that for the improvement in forecasting and classification, the forecast horizon is a good feature in the task for the model selection. MRMR (Minimum Redundancy Maximum Relevance) can improve the model choice framework's performance even further, implying that a viable feature reduction strategy should be used in version selection to increase classifier generalization capabilities. Finally, distinctive distributions of time series capabilities may result in different forecast styles. It demonstrates the importance of designing a primarily model based on time series characteristics. In Anggraeni et al. (2017), the authors proposed ARIMAX model and VAR (Vector Autoregressive) model used for the prediction of rice price concerning Indonesian society. ARIMAX model with an MAPE of 0.15% performed better than VAR model with an MAPE of 15.27%. In Xu and Zhang (2021), the authors used a univariate and bivariate neural network (NN) modeling for the forecasting price of corn cash. In Weng et al. (2019), the authors applied RNN (Recurrent Neural Network) and ARIMA techniques for the prediction for the price of horticulture products and obtained an average relative error of 10.70%.

7.3. Materials and methods

7.3.1. *Data source*

Rice production time-series data is taken in the presented work. It is taken from "indiastat.com" (see: https://www.indiastat.com/table/rice/season-wise-area-production-productivity-rice-indi/7264).

7.3.2. *Data description*

The collected data contains two columns, the first is for the session/year and the second is for rice production in 1,000 tons. The size of the dataset is (73 × 2). The given data is for 73 years, i.e. from 1949 to 2021. The structure of the dataset is shown in Figure 7.4.

	Session	Production_in_1000_tonne
0	1949	23500
1	1950	20576
2	1951	21300
3	1952	22899
4	1953	28214
...	...	...
68	2017	112758
69	2018	116478
70	2019	118870
71	2020	122270
72	2021	104300

73 rows × 2 columns

Figure 7.4. *Data structure*

7.3.3. *Coding environment*

Configuration of the experimental setup is given in Table 7.2.

Processor	Intel(R) Core(TM) i3-6100 CPU @3.70GHz
RAM	8.00 GB
System type	64-bit operating system, x64-based processor
Windows edition	Windows 10 Education
Windows version	21H2
Python version	3.8.12
Computing platform	Jupyter-Notebook

Table 7.2. *Experimental setup configuration*

Classification of forecasting techniques

Based on the customers and expert opinion and historical data of demand, price and production estimation or prediction for the future using scientific calculation are known as forecasting (Mahto et al. 2021). The classification of the forecasting is shown in Figure 7.5.

7.3.3.1. *Subjective techniques*

In this type of prediction, market analysis and expert opinions are taken. These are purely theoretical in nature and no numerical values are used.

7.3.3.2. *Numerical techniques*

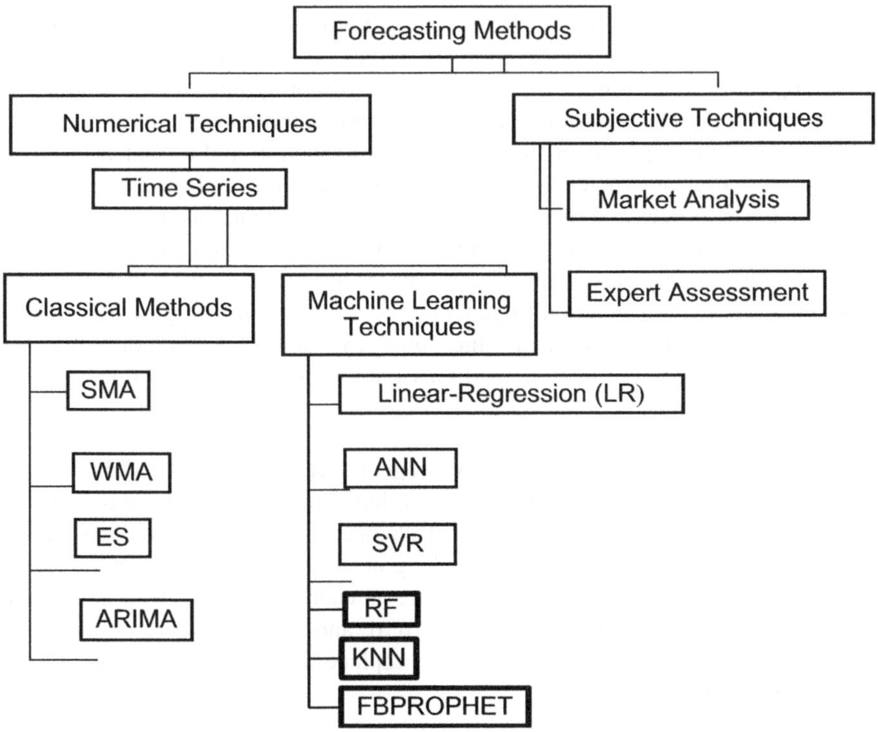

Figure 7.5. *Classification of forecasting techniques*

In this forecasting method, recent data is used for the prediction of future values by using classical numerical methods and ML techniques. SMA is a very fundamental technique, in this, the forecasting can be done by finding the normal mean of previous data. In weighted moving average (WMA), recent data are not taken of the same weightage. The very latest data is taken of higher weightage and consequently, old data is taken of minimum weightage. Exponentially smoothed (ES) forecasts are weighted averages of previous observations, with the weights decaying exponentially as the observations get older. LR is a basic statistical regression method that demonstrates the relationship across continuous data which is used for predictive modeling. In LR, input variable (x) is the dependent variable on which output depends. This becomes in terms of y = ax + b, where a and b are the slope and intercept, respectively. In our study, the year is taken as the input and rice product is considered as the output.

Methods used in this study

Artificial neural network (ANN) (Mohamed 2019): an ANN is a computing system that tries to mimic a human brain. The ANN model is a type of intelligent system that can be used to tackle complex issues in a variety of applications, including optimization, prediction, modeling, grouping, pattern recognition, simulation and more. The ANN structure is made up of three layers: an input layer that collects data, an output layer that computes information and a hidden layer as well as one or more hidden levels for connecting the input and output layers. A neuron is a basic processing unit in a neural network that has two functions: collecting data and processing it. The inputs and the result production connection weights are multiplied by each input, and its biases are added, after which they are sent through an activation function to generate the output seen in Figure 7.6. In our study, look back function is used in which one-year data of production works as input whereas next year production works as output. Similarly, it is done for all available data.

Autoregressive integrated moving average (ARIMA) model (Mahto et al. 2021): ARIMA is a type of regression analysis that shows how strong a dependent variable is in relation to other variables that change. The model's final goal is to forecast future time series movement by looking at discrepancies between values in the series rather than actual values. ARIMA treats each component as a parameter with a consistent nomenclature. ARIMA with p, d, and q is a standard notation for ARIMA models, where values of the variables replace the parameters to denote the kind of ARIMA model used.

p: the model's lag order, sometimes known as the number of lag observations. *d*: the degree of differencing and is also known as the number of times the raw observations are differed. *q* is the size of the moving average window and is also known as the moving average order.

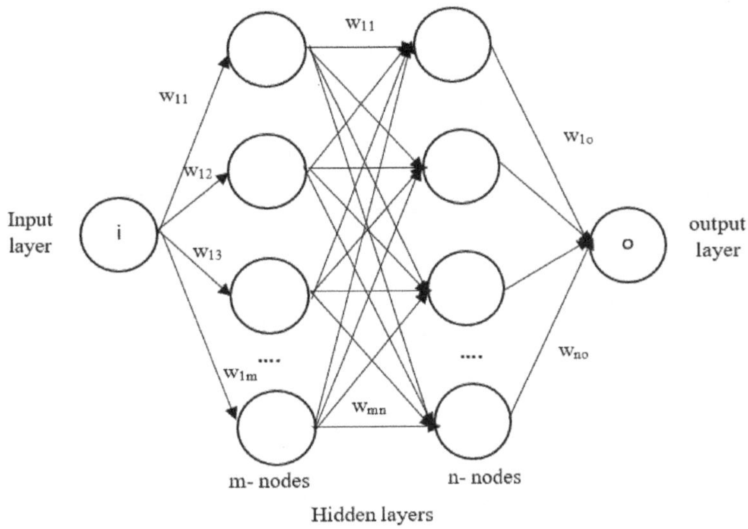

Figure 7.6. *ANN model*

Data size	(73, 2)
Training size	(60, 2)
Testing size	(11, 2)
Look-back size	2
Number of epochs	50
Batch_size	40
Input layer	Look back
Number of hidden layers	2
Nodes at the first hidden layer	32
Nodes at the second hidden layer	8
Activation function	Look back

Table 7.3. *ANN parameters used in the experiment*

Support vector regressor (SVR) model (Cantillo-Luna et al. 2022): the idea behind the SVR is similar to the SVM (Support vector machine) where data is classified by fitting the hyperplane. In the SVR, a hyperplane is created for the given data having the maximum number of data points. This plane is similar to the line of best fit. In the present experiment, data is firstly split into training (85%) and testing (15%) datasets. For the prediction, sklearn library with linear kernel is used.

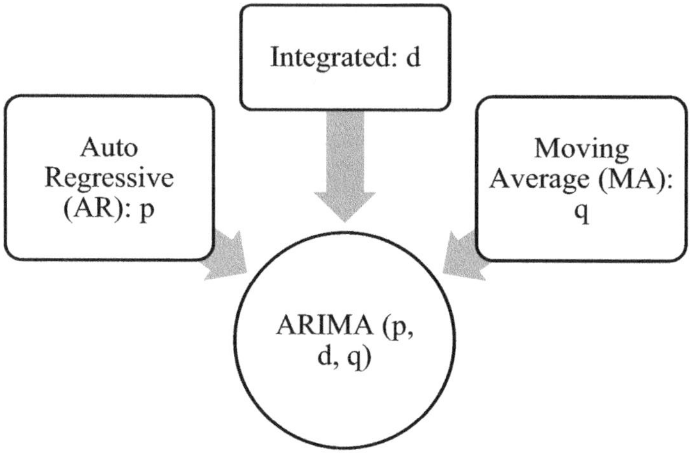

Figure 7.7. *ARIMA model*

Random forest regressor (RF) model (Gupta et al. 2022): RF is a machine learning technique that can be used as a regression model. Additionally, RF naturally and data-drivenly captures any potential interactions between the factors on which the predicted value depends. This algorithm also works on the nonlinear dataset. Because of economic theory predicts that the impact of uncertainty on financial factors is likely to be nonlinear, dealing for such nonlinearities may be significant. Additionally, the calculation of random forests is governed by an easy-to-understand method, which is unquestionably advantageous considering that random forests have not yet been thoroughly researched by the economics profession.

k-nearest neighbors (KNN) model (Li and Jin 2022): one of the most straightforward techniques in data analysis classification technology is the neighbor algorithm. Each sample is said to be represented by its k closest neighbors. The kNN algorithm's core tenet is that if the majority of the k closest samples in a feature space fall into a certain category and the sample itself does too and shares the traits of samples in this category. When making a classification decision, the kNN approach only considers the category of one or more nearby samples when categorizing a sample. In the present study, we used five numbers of neighbors.

FB-Prophet model (Saiktishna et al. 2022): this model is used by Facebook for analyzing their time-series data. The input of the data is given in the format of

YYYY-MM-DD HH:MM:SS. In this Python library, the following nomenclature is used:

ds = date input, yhat = mean value of prediction, yhat_lower = lower value of prediction, yhat_upper = upper value of prediction.

The structure results obtained by using this model are shown in Figure 7.8.

	ds	yhat	yhat_lower	yhat_upper
0	1949-01-01	20154.307961	10854.492443	29218.274222
1	1950-01-01	20415.864263	11409.007714	29787.639861
2	1951-01-01	21585.429919	11887.452497	31344.051017
3	1952-01-01	23662.405678	15026.116425	32423.224612
4	1953-01-01	24283.534310	14418.551142	33731.833550
...	...	...	...	...
69	2018-01-01	111607.090747	102273.303668	121251.884535
70	2019-01-01	113276.081849	103954.603922	122147.696859
71	2020-01-01	115852.483051	106621.738811	124913.207899
72	2021-01-01	116974.405415	107447.246600	126335.190567
73	2021-12-31	123717.162958	114309.231062	132913.087445

Figure 7.8. *FB-Prophet output*

Performance metrics

7.3.4. *Forecasting evaluation*

The following relations are used in the present study to find the accuracy or error in the prediction:

$$MAPE = \frac{1}{n}\sum_{i=1}^{n}\left|\frac{y_{pi} - y_i}{y_i}\right| * 100\ \%$$

$$RMSE = \sqrt{\sum_{i=1}^{n} \frac{\left(y_{pi} - y_i\right)^2}{n}}$$

$$MAE = \sum_{i=1}^{n} \frac{|y_{pi} - y_i|}{n}$$

where y_{pi} and y_i are the predicted and actual production quantity in 1,000 of tons, respectively for the ith year and n = number of years

Workflow of the proposed work

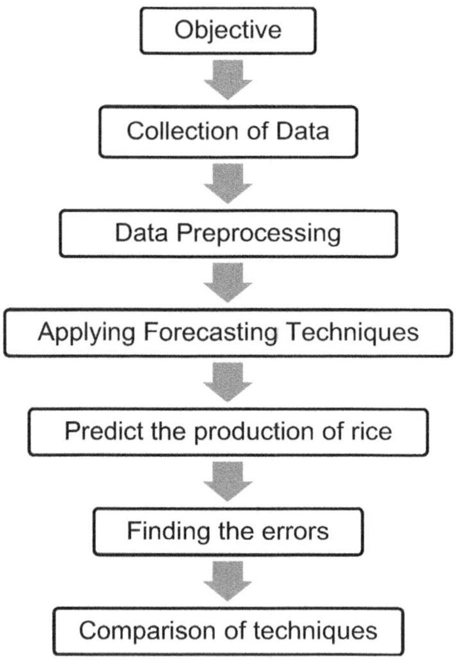

7.4. Results and discussion

The trend of the production of rice (in 1,000 tons) in India is shown in Figure 7.12 between 1949 and 2021. The increase in the production indicates that new technologies, good qualities of seeds are being used in the agriculture. This is also necessary as the population is continuously growing. The result of forecasting for SMA, WMA, ES, ARIMA, LR, ANN, SVR, RF, KNN and FB-Prophet is demonstrated in Table 7.2 and Figure 7.12, which indicates that the trending of production in future will continue to rise. With the help of this prediction, we can also estimate the requirement of the capacity for storage, export capacity.

The accuracy of applied techniques is demonstrated in Table 7.2, and the errors, i.e. MAPE, RMSE and MAE, are shown in Figure 7.13. RF is performed very well for the used data of rice in respect of accuracy and in terms of execution time.

For rice production time-series data, using the RF model, the MAPE value is 2.71%. However, using SMA, WMA, ES, ARIMA, LR, ANN, SVR, KNN and FB-Prophet models, the MAPE values are 8.95%, 8.48%, 11.08%, 5.03%, 9.13%, 4.32%, 12.12%, 5.44% and 6.26%, respectively.

Similarly, RMSE values for SMA, WMA, ES, ARIMA, LR, ANN, SVR, KNN and FB-Prophet models are 6281.82, 5960.65, 8041.47, 6102.74, 5615.92, 5189.31, 9885.09, 3918.44 and 4678.87, respectively, and MAE values for SMA, WMA, ES, ARIMA, LR, ANN, SVR, KNN and FB-Prophet models are 5385.35, 5060.93, 6934.77, 5348.28, 4495.39, 3878.50, 7866.08, 2959.85 and 3503.25, respectively.

S. no.	Model	MAPE	RMSE	MAE	Forecasting for 2022 (in 1,000 tons)
1.	SMA (four years)	8.95	6281.82	5385.35	115479.5
2.	WMA (four years)	8.48	5960.65	5060.93	113822.8
3.	ES ($\alpha = 0.2$)	11.08	8041.47	6934.77	110410.7
4.	ARIMA	5.03	6102.74	5348.28	113022.9
5.	LR	9.13	5615.92	4495.39	114973.7
6.	ANN	4.32	5189.31	3878.50	117319.9
7.	SVR	12.12	9885.09	7866.08	98611.5
8.	RF	2.71	1977.79	1500.35	108892.7
9.	KNN	5.44	3918.44	2959.85	114935.2
10.	FB-Prophet	6.26	4678.87	3503.25	123717.1

Table 7.4. *Results and evaluation of forecasting techniques*

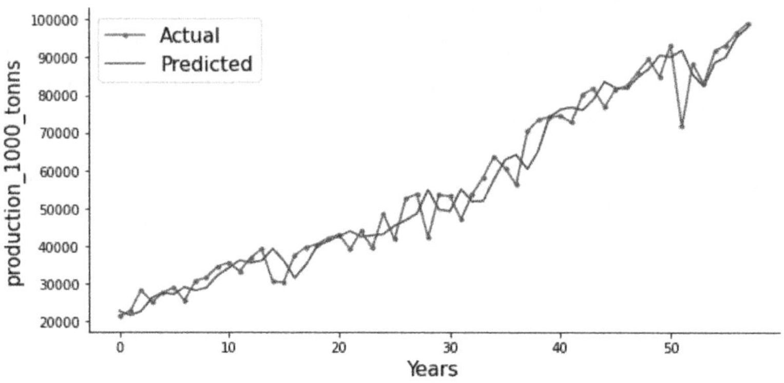

Figure 7.9. *Forecasting using ANN. For a color version of this figure, see www.iste.co.uk/koubaa/optimization.zip*

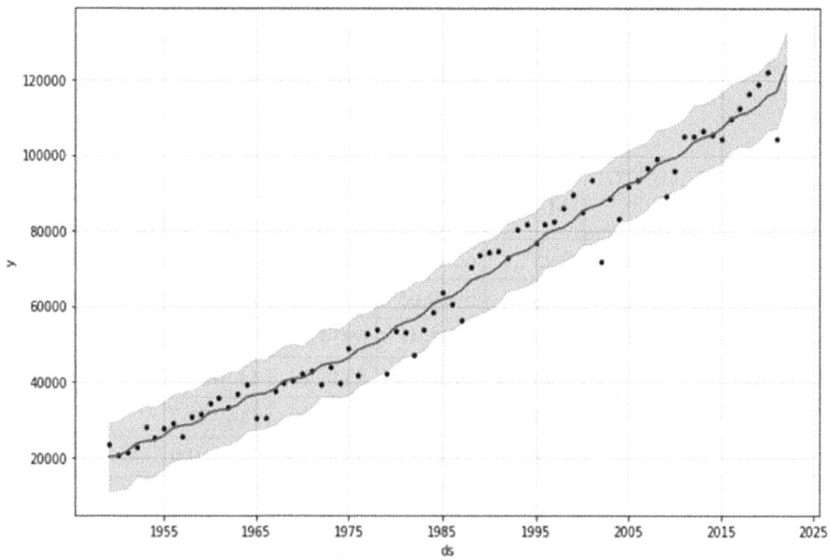

Figure 7.10. *Forecasting using FB-Prophet. For a color version of this figure, see www.iste.co.uk/koubaa/optimization.zip*

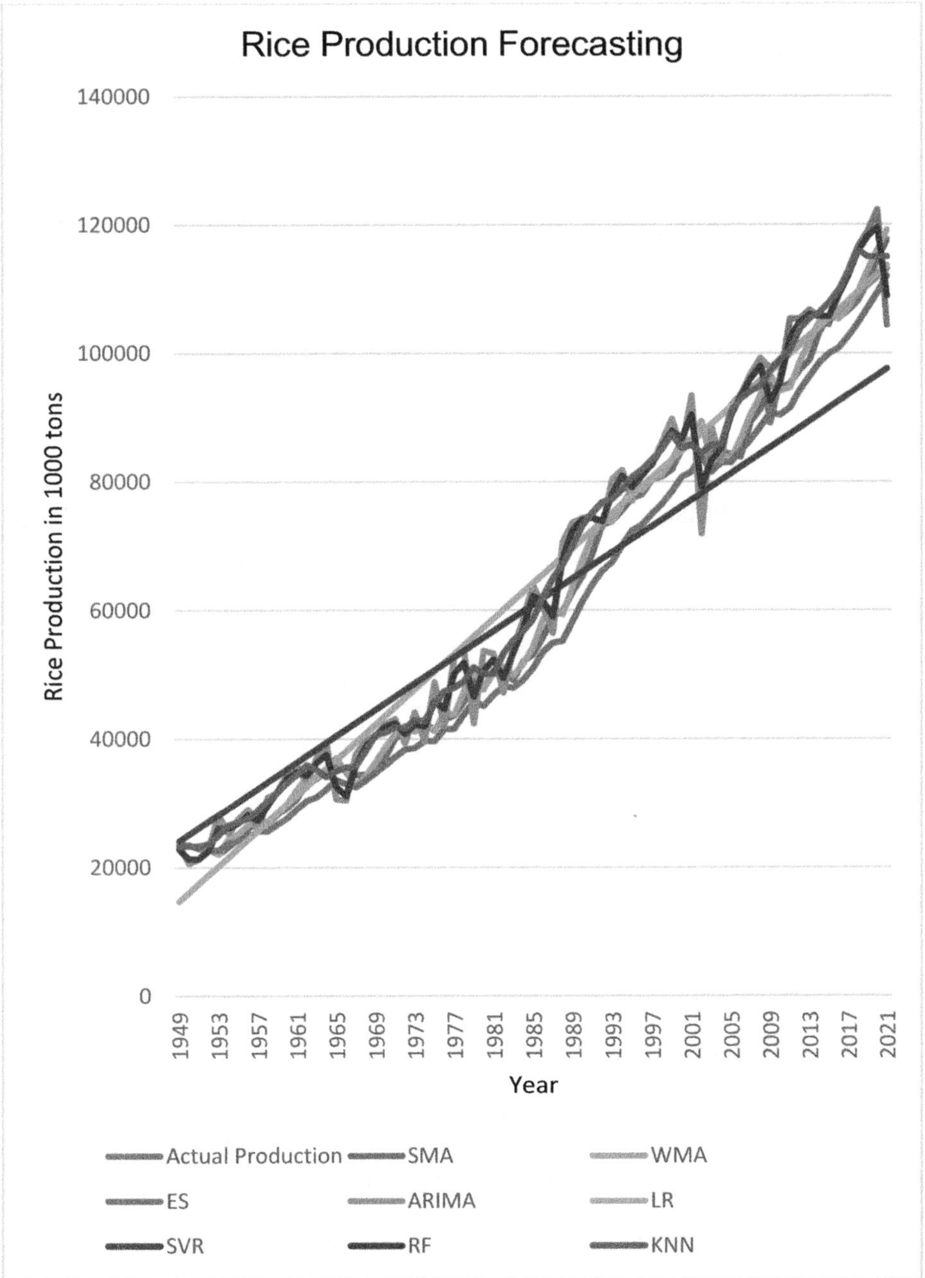

Figure 7.11. *Results of forecasting by different models. For a color version of this figure, see www.iste.co.uk/koubaa/optimization.zip*

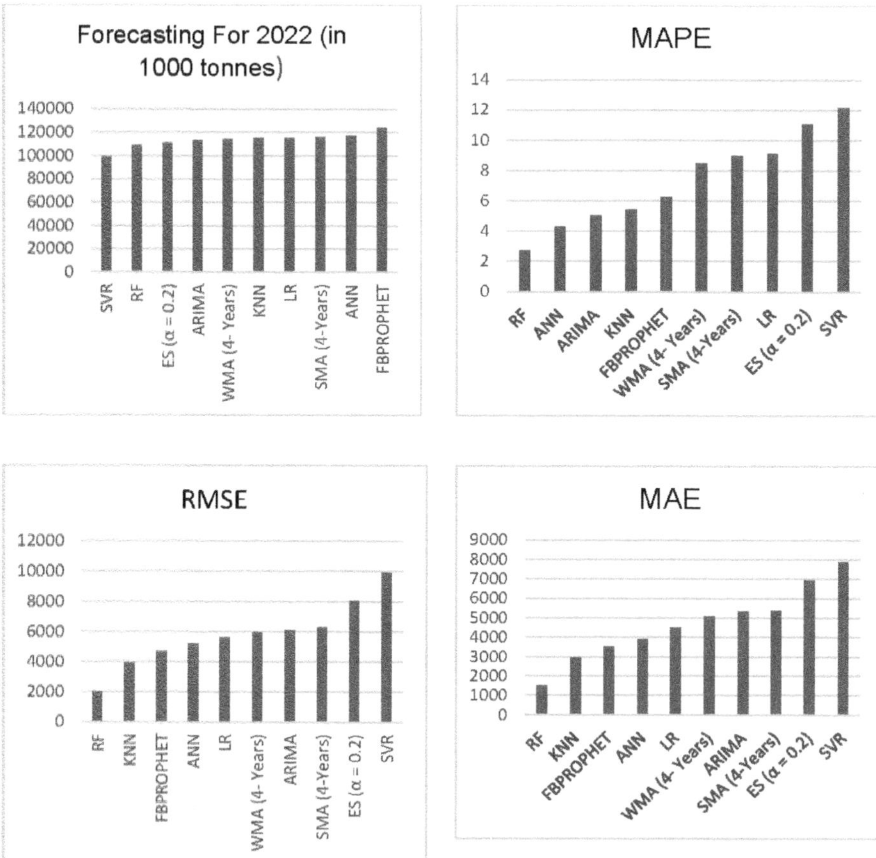

Figure 7.12. *Forecasting production for 2022 and model errors. For a color version of this figure, see www.iste.co.uk/koubaa/optimization.zip*

7.5. Conclusion and future scope

India is currently ranked the second in the world for agricultural commodity output and accounts for over 18% of the country's GDP. Considering the population growth in India, it is necessary to predict the production volume of AFP to meet the demand of the consumers. The focus of the present study is to forecast the production of rice using classical as well as ML techniques. For this, 10 forecasting methods including SMA, WMA, ES, ARIMA, LR, ANN, SVR, RF, KNN and FB-Prophet techniques are employed and are compared using the performance metrics as: MAPE, RMSPE and MAE. The following observations were made from the given data:

1) Analyzing the performance of ML and classical techniques, it can be observed that for ML algorithms the error is considerably lower in comparison to the classical techniques. SVR is an exception here but its results can be improved on increasing the size of data (like monthly data).

2) When the data size increases, the execution time is likely to be more for classical techniques in comparison to ML techniques.

3) For rice production time series data, maximum and minimum values of MAPE are 12.12% (SVR) and 4.32% (RF), respectively.

4) Similarly, maximum and minimum values of RMSE are 9885.09 (SVR) and 1977.79 (RF), respectively. Also, the maximum and minimum values of MAE are 7866.08 (SVR) and 1500.35 (RF), respectively.

5) The RF model performed better than the other forecasting methods for the given data.

6) The classical models provide the result using the analysis of very previous data. However, ML models analyze the whole data at a time which is helpful in finding the trend of production. This can also be a reason for better results of ML models.

7) In this study, a simple model was considered for forecasting the production of rice. Several factors that affect the production of the crop may be considered as part of the future study to make more accurate forecasting.

8) The problem can be made more realistic and challenging by considering the data in terms of date and time (rather than years, as is done in the present study). This may also help enhance the results of FB-Prophet.

7.6. References

Anggraeni, W., Andri, K.B., Sumaryanto, M.F. (2017). The performance of ARIMAX model and vector autoregressive (VAR) model in forecasting strategic commodity price in Indonesia. *Procedia Comput. Sci.*, 124, 189–196.

Banerjee, T., Sinha, S., Choudhury, P. (2022). Long-term and short-term forecasting of horticultural produce based on the LSTM network model. *Appl. Intell.*, 52, 9117–9147.

Cantillo-Luna, S., Moreno-Chuquen, R., Chamorro, H.R., Riquelme-Dominguez, J.M., Gonzalez-Longatt, F. (2022). Locational marginal price forecasting using SVR-based multi-output regression in electricity markets. *Energies*, 15(1), 293.

Das, A., Senapati, M., John, J. (2009). Impact of agricultural credit on agriculture production: An empirical analysis in India. *Reserve Bank India Occas. Pap.*, 30(2), 75–107.

Gupta, R., Pierdzioch, C., Salisu, A.A. (2022) Oil-price uncertainty and the U.K. unemployment rate: A forecasting experiment with random forests using 150 years of data. *Resour. Policy*, 77, 102662.

Hellerstein, D. and Scott, M. (2011). The influence of rising commodity prices on the conservation reserve program. *Econ. Res. Serv.*, 110.

Li, F. and Jin, G. (2022). Research on power energy load forecasting method based on KNN. *Int. J. Ambient Energy*, 43(1), 946–951.

Li, B., Ding, J., Yin, Z., Li, K., Zhao, X., Zhang, L. (2021). Optimized neural network combined model based on the induced ordered weighted averaging operator for vegetable price forecasting. *Expert Syst. Appl.*, 168, 114232.

Mahto, A.K., Alam, M.A., Biswas, R., Ahmed, J., Alam, S.I. (2021). Short-term forecasting of agriculture commodities in context of indian market for sustainable agriculture by using the artificial neural network. *J. Food Qual.*, 1–13.

Mohamed, Z.E. (2019). Using the artificial neural networks for prediction and validating solar radiation. *J. Egypt. Math.*, 27(1), 1–13.

Reserve Bank of India (2021). *Handbook of Statistics on Indian States*. Reserve Bank of India, Mumbai.

Ribeiro, M.H.D.M. and dos Santos Coelho, L. (2020). Ensemble approach based on bagging, boosting and stacking for short-term prediction in agribusiness time series. *Appl. Soft Comput.*, 86, 105837.

Sabu, K.M. and Kumar, T.K.M. (2020). Predictive analytics in agriculture: Forecasting prices of Arecanuts in Kerala. *Procedia Comput. Sci.*, 171, 699–708.

Saiktishna, C., Sumanth, N.S.V., Rao, M.M.S., Thangakumar, T. (2022). Historical analysis and time series forecasting of stock market using FB prophet. In *2022 6th Int. Conf. Intell. Comput. Control Syst. ICICCS*. IEEE, Madurai, 1846–1851.

Weng, Y., Wang, X., Hua, J., Wang, H., Kang, M., Wang, F.-Y. (2019). Forecasting horticultural products price using ARIMA model and neural network based on a large-scale data set collected by web crawler. *IEEE Trans. Comput. Soc. Syst.*, 6(3), 547–553.

Xu, X. and Zhang, Y. (2021). Corn cash price forecasting with neural networks. *Comput. Electron. Agric.*, 184, 106120.

Xu, Y., Shen, S., Chen, Z. (2009) Research on forecast of sugar price based on improved neural network. In *2009 Second Int. Symp. Intell. Inf. Technol. Secur. Inform.* IEEE, Moscow, 12–15.

Yadav, V.S., Singh, A.R., Gunasekaran, A., Raut, R.D., Narkhede, B.E. (2022) A systematic literature review of the agro-food supply chain: Challenges, network design, and performance measurement perspectives. *Sustain. Prod. Consum.*, 29, 685–704.

Zhang, D., Chen, S., Liwen, L., Xia, Q. (2020). Forecasting agricultural commodity prices using model selection framework with time series features and forecast horizons. *IEEE Access*, 8, 28197–28209.

8

Prioritization of Circular Olive Oil Production in Tunisia based on Multi-Criteria Decision-Making within a Sustainable Life Cycle Framework

8.1. Introduction

Olive oil production has risen steadily worldwide, from 1.8 million tons per year in the 1990s to 3 million tons per year in 2017 (Rapa and Ciano 2022). There was a 12% increase on the previous year's production. For the 2019 season, Spain ranked as the world's leading olive oil producer[1] with an average of 1,738,600 tons. Italy came second with 294,914 tons. The other main producers were Greece with 208,900 tons and Tunisia with 179,700 tons (source: Commission Européenne 2019). 98% of the world's olive population is located in the Mediterranean basin.

Despite the economic importance of this food product in many countries, the olive oil industry demonstrates environmental impacts in terms of resource depletion, land degradation, air emissions and waste production (Duman et al. 2020). These impacts can vary considerably due to the practices and techniques used in olive cultivation and oil production. These vary greatly from one country to another, as well as from one region to another within the same country, depending on local plantation characteristics and management practices, as well as socio-economic, climatic and sometimes even cultural conditions (Duman et al. 2020).

Chapter written by Mohamed Amir KESKES, Alaeddine ZOUARI, Rémy HOUSSIN, Diala DHOUIB and Jean RENAUD.

1 Olive oil is a type of edible oil, made from vegetable fat extracted from olives (the fruit of olive trees grown in olive groves) during crushing in an oil mill.

The aim of this chapter is to help decision-makers, within a framework of sustainable development, to choose the most sustainable agricultural practice, processing method and waste recovery management, so that all stages in the life cycle of the olive oil chain are circular[2] and sustainable. Therefore, by applying the principles of the circular economy, we can identify the different life cycle scenarios of the olive oil processing chain. By implementing the environmental life cycle assessment (E-LCA), the life cycle cost analysis (LCCA) and the social life cycle assessment (S-LCA), it is possible to carry out a life cycle sustainability assessment (LCSA[3]) for each scenario related to the olive oil processing chain.

In this way, it is possible to compare different scenarios with the aim of helping stakeholders to decide on their choice of strategy using a multi-criteria decision support framework. Our study evaluates the sustainability of existing olive-growing systems in Tunisia, based on eight alternatives proposed in the scientific literature.

8.2. Literature review

The life cycle sustainability assessment (LCSA) is used to evaluate the environmental, economic and social impacts of production systems (Fernández-Tirado et al. 2021). However, assessing these impacts is not always easy. For this reason, decision-making must be carried out within the framework of multi-criteria decision analysis (MCDA[4]). For example, Ren and Toniolo (2018) have developed a methodological framework for multi-criteria decision-making (MCDM) in the life cycle of sustainable bioethanol production. These authors assigned weights for the different criteria using the AHP[5] method, then applied the VIKOR[6] method to determine the most sustainable sequence of alternatives.

The AHP tool was also used by Foolmaun and Ramjeawon (2013) with LCSA to integrate the three dimensions of sustainability (environmental, economic and social). In this study, a comparison was made between different disposal scenarios for used polyethylene terephthalate bottles, to obtain the most sustainable scenario for this type of waste. In this context, the AHP tool was also used to assign weights to the three dimensions of sustainability and to sort the scenarios. Zhang and

2 According to the *Agence de la Transition Ecologique*, "the circular economy is defined as an economic system of trade and production which, at all stages in the life cycle of products (goods and services), aims to increase the efficiency of resource use and reduce the impact on the environment, while enhancing the well-being of individuals".

3 LCSA = E-LCA + LCCA + S-LCA.

4 MCDA: multi-criteria decision analysis.

5 AHP: analytic hierarchy process (multi-criteria method).

6 VIKOR: *VlseKriterijuska Optimizacija I Komoromisno Resenje* (multi-criteria method).

Haapala (2015) also used MCDA tools in the context of LCSA to enable comparison between different scenarios, based on a multi-criteria model. The authors specifically used the AHP tool for assigning weights and the PROMETHEE[7] method for ranking alternatives.

In the context of olive oil, Riesgo and Gallego-Ayala (2015) conducted a study aimed at measuring the sustainability of olive farms in the first instance. They then identified the most sustainable olive-growing system in Spain. In a second step, the authors (Jellali et al. 2021) presented a multi-criteria approach in a "fuzzy" context, and evaluated different scenarios for the agricultural phase and the olive oil processing phase in Italy. The authors used the TOPSIS method, which gives approximate values.

This literature review shows that the combination of life cycle sustainability and multi-criteria tools is often used by several authors and in several fields. However, few articles deal with the problems of olive oil life cycle sustainability. The contribution of this work is to propose a better configuration of the Tunisian olive oil supply chain from a sustainable development perspective. The proposed approach involves combining the best worst method (BWM), similar to the AHP, to determine the weight of sustainability evaluation criteria, and the VIKOR method for ranking life cycle scenarios, based on a set of criteria.

8.3. Proposed methodology

The methodology used in our study consists of three stages, which is shown in Figure 8.1.

– Step 1 includes a review of the literature on olive oil farming practices, waste management, studies and analyses in the Sfax region (Tunisia), based on surveys of decision-makers[8]. The aim of this stage is to extract eight alternatives and the corresponding criteria. For the alternatives, the SADT method[9] was integrated into our approach to gain a broader vision through a top-down analysis of successive levels. This method enables us to specify the role of each of the system's elements in greater detail, and to specify the feedback loops for each scenario. The result is the construction of sustainable scenarios within the framework of the circular economy. The criteria used in this study (E-LCA, LCCA and S-LCA) were validated by

7 PROMETHEE: Preference Ranking Organization Method For Enrichment Evaluation.

8 Stakeholders in the olive-growing sector in the city of Sfax (farmers, olive mills, Institut de l'Olivier).

9 SADT (Structured Analysis and Design Technique) is a graphical method that starts with the general and works down to the specific. It is used to describe systems.

specialist experts (Sfax farmer, Institut de l'olivier Sfax, etc.) in the agricultural field.

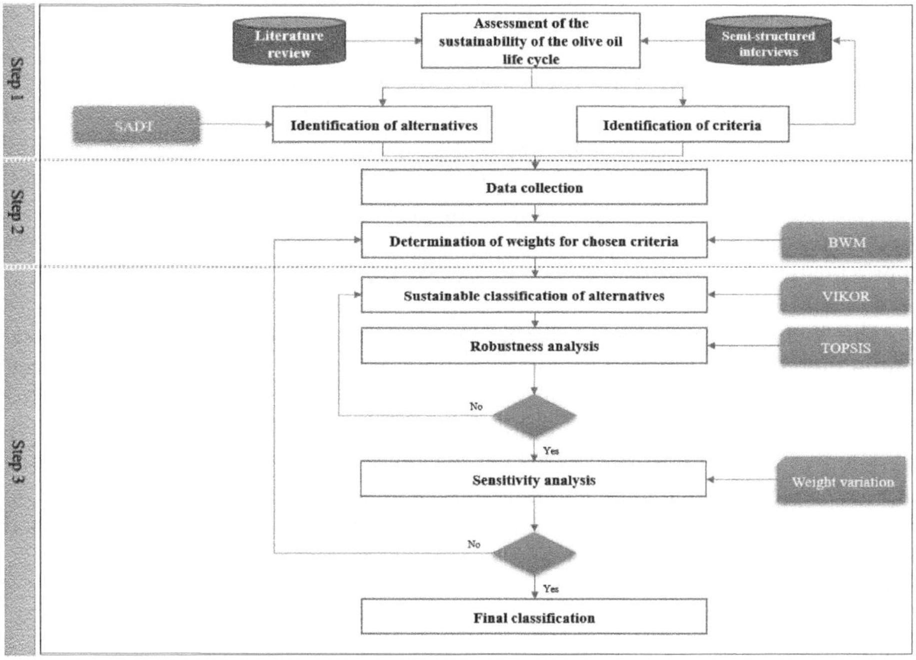

Figure 8.1. *Methodological framework for selecting the most sustainable scenario. For a color version of this figure, see www.iste.co.uk/koubaa/optimization.zip*

– Step 2 integrates data collection from surveys with decision-makers to build the decision matrix. Then, the weight values of the sustainability evaluation criteria are determined using the BWM. This has many advantages over the MCDM approach, such as the AHP (Rezaei 2015). Due to fewer pairwise comparisons, BWM saves data collection, calculation and analysis time compared to the AHP. BWM provides results with better reliability and minimal inconsistency.

– Step 3 consists of sequencing the different olive oil life cycle scenarios using the VIKOR method. The advantage of this method is that data can be treated as qualitative, quantitative or with different units of measurement in the decision matrix. To reduce the impact of uncertainties and human-induced disturbances, we performed a sensitivity analysis by changing the criteria weights. This ensured the stability of the result obtained. Finally, the TOPSIS method was used to check the robustness of the results of the VIKOR calculation and to analyze the consistency of the results.

8.3.1. *Calculation of criteria weights using the BWM*

The BWM[10] comprises five steps, according to Rezaei (2015):

– Step 1: to determine a set of decision criteria.

– Step 2: to determine the best (e.g. most desirable, most important) and worst (e.g. least desirable, least important) criteria for the decision-maker. In this step, no comparison is made at this stage.

– Step 3: to determine the preference of the best criterion over all other criteria, using a number between 1 and 9. The best vector would be: $A_B = (a_{B1,}\ a_{B2,}\ a_{Bn})$, where $a_{Bj,}$ indicates the preference of the best criterion B over criterion j. Clearly $a_{BB} = 1$.

– Step 4: to determine the preference of all criteria over the worst criterion, using a number between 1 and 9. The resulting Other to Worst vector would be: $\mathrm{A_W} = (\mathrm{a_{W1,}\ a_{W2,}\ a_{nW}})^{\mathrm{T}}$, where $\mathrm{a_{jW,}}$ indicates the preference of criterion *j* over the worst criterion *W*. Clearly, $a_{WW} = 1$.

– Step 5: to find the optimum weights $(W_1^*, W_2^*, \ldots\ldots, W_n^*)$.

The problem can be summarized to the next mathematical model:

Min ξ	
c/o	W_B best B criterion on J,
$\lvert W_B - a_{BJ} W_J \rvert \leq \xi$, for all j [8.1]	W_J the worst criterion on W,
$\lvert W_j - a_{JW} W_W \rvert \leq \xi$, for all j [8.2]	a_{BJ} preference of best criterion B over criterion J,
$\sum_j W_J = 1$ [8.3]	
$W_J \geq 0$, for all j. [8.4]	a_{JW} the preference of criterion j over the worst-case criterion W.

By solving the problem, the optimal weights $(W_1^*, W_2^*, \ldots\ldots, W_j^*)$ and ξ^* are obtained.

10 BWM is based on a systematic pairwise comparison of decision criteria.

8.3.2. *Ranking of alternatives using the VIKOR method*

The stages of the VIKOR methodology are described below:

– Step 1: construction and standardization of the decision matrix: the quantitative values of the selected criteria (i.e. environmental, economic and social indicators) are expressed in matrix form, and their various units of measurement are homogenized by linear standardization to produce dimensionless indices:

Cost normalization method: *Minimization*

$$x_{ij} = \frac{f_{max} - f_{ij}}{f_{max} - f_{min}} \quad [8.5]$$

Benefit normalization method: *Maximization*

$$x_{ij} = \frac{f_{ij} - f_{min}}{f_{max} - f_{min}} \quad [8.6]$$

– Step 2: to determine the best (f_j^*) and worst (f_j^-) values for each *Cj;* therefore, equations [8.7] and [8.8] are used as follows:

If the criterion function is a benefit

$$f_j^* = max f_{ij}; f_j^- = min f_{ij} \quad [8.7]$$

If the criterion function is a cost

$$f_j^* = min f_{ij}; f_j^- = max f_{ij} \quad [8.8]$$

– Step 3: in order to work out the distance of the alternatives from the ideal solution, we need to calculate the indices S_k equation [8.9] and R_K equation [8.10]. The first S_k represents the distance of an alternative to a positive ideal solution, while the second R_K represents the distance of an alternative to a negative ideal solution:

$$S_i = L_i^{p=1} = \frac{\sum_{j=1}^{n} w_j^* \left(f_j^* - f_{i,j}\right)}{\left(f_j^* - f_j^-\right)} \quad [8.9]$$

$$R_k = L_i^{p=\infty} = max_j \left[W_j^* \frac{\left(f_j^* - f_j^-\right)}{\left(f_j^* - f_j^-\right)} \right] \quad [8.10]$$

– Step 4: we then calculate Q_i (equation [8.11]), which represents the VIKOR synthetic index. The lowest value of Q_K corresponds to the best scenario, which is closest to the ideal value, ϑ represents the weight of the alternative with the highest utility, and is normally equal to 0.5:

$$Q_i = \vartheta \frac{(S_i - S^*)}{(S^- - S^*)} + (1 - \vartheta) \frac{(R_i - R^*)}{(R^- - R^*)} \quad [8.11]$$

– Step 5: when $\vartheta=0.5$➔, both conditions must be met:

- *Condition 1*: $(Q(A_2) - Q(A_1) \geq (1/(n-1))$ [8.12]

where n is the number of alternatives, A_1 is the best alternative in the ranking list, and A_2 is the second best alternative in the ranking list.

- *Condition 2*: A_1 must also be the best classified by S or/and R.

Once these two conditions have been met, VIKOR can rank the alternatives to quickly and accurately determine the best scenario.

8.4. Olive oil production and waste management case study

In order to illustrate our MCDA methodology, presented in Figure 8.1, for decision-making on the selection of the best scenario for the olive oil life cycle in the particular context of Tunisia, a concrete illustrative case is presented. This case concerns the most sustainable agricultural practices, such as the processing method and waste recovery management.

8.4.1. *The olive oil life cycle*

The study covers the following three phases: olive production, processing and waste management (Appendix 1).

8.4.1.1. *Phase 1. Olive production, farming and cultivation*

The agricultural phase includes the main farming practices applied by farmers (Appendix 1), which are as follows:

– *Irrigation* affects olive yield, fruit size, quality characteristics and ripening time. Given the lack of water in the region, olive trees are generally grown without irrigation (dry method). Some farmers use drip irrigation to improve productivity.

– *Soil management* includes all operations required for weeding and levelling the site.

– *Pruning and treatment of residues* are carried out by the pruners once a year, using appropriate chainsaws. Brush can be crushed or burned. Grinding residues or ashes are then scattered over the fields.

– *Harvesting* can be carried out mechanically (vibrating comb, rotary comb, mechanical harvesters) or manually (traditional method with workers).

– *The fertilization process* can vary greatly due to the use of different materials/substances (NPK fertilizers, direct application of mill wastewater to the soil, or with wet pomace manure and compost), and due to the method of distribution: mechanical (with tractors) or manual (with olive trees).

– *Pest control* in conventional agriculture is generally carried out manually, with workers using pesticides twice a year. Generally, in Tunisia, climatic conditions and farming practices reduce the use of pesticides, so that olive trees are grown naturally without treatment, especially in organic fields, where only the use of baits is included.

8.4.1.2. *Phase 2. Production and processing of virgin olive oil*

After harvesting, the olives are quickly brought to the mills in bulk and transformed into olive oil, within 24 hours to avoid fermentation phenomena (Salomone and Ioppolo 2012). There are three olive oil processing systems: traditional pressure system, two-phase centrifugal system and three-phase centrifugal system:

– *The discontinuous system* corresponds to traditional pressing, generally used in certain small traditional mills (*made up of stone wheels on a granite slab*). The olives are crushed to a paste between straw disks under high pressure. Finally, the oil is centrifuged to remove water and residual solids.

– *Continuous centrifugation with a three-phase system*, while having a higher production capacity than traditional pressing, has certain disadvantages, such as higher water and energy consumption (*due to the addition of hot water to dilute the olive paste*). This process uses a three-phase decanter that generates solid waste (pomace) and liquid waste (wastewater) and olive oil.

– *Continuous centrifugation with a two-phase system* that separates the oil from the olive paste without adding water, thus eliminating the problem of vegetation water. It combines the advantages of the two different systems, as it requires the addition of a small quantity of water and generates a solid fraction that includes part of the vegetation water, pomace, and a smaller quantity of wastewater.

8.4.1.3. *Phase 3. Waste management*

The waste treatment (by-product) phase includes all processes related to the treatment of solid and liquid fractions from olive oil production (Appendix 1). It should be noted that olive oil represents only 20% of the olive mass. The remaining 80% is waste. Consequently, olive oil processing is characterized by the generation of huge quantities of wastewater and pomace.

Pomace varies according to the olive crushing system used. In fact, the waste resulting from a discontinuous system and a continuous three-phase system is dry pomace. In the case of a continuous two-phase system, it is wet pomace. The management of these two types of sludge differs.

For dried pomace, after a drying process, the oil is extracted using a solvent (hexane). This process produces pomace oil, as well as spent pomace and pulp, which is mainly used as fuel, since dried olive pomace has a high calorific value.

For wet pomace processing, the pulp and kernel are removed. The pulp is used to produce compost, and the kernel is sold as fuel and for animal feed.

Mill wastewater is the liquid effluent generated by the crushing of olives to obtain oil. It represents a real environmental problem for producing countries. Mill wastewater is generally discharged into basins throughout Tunisia. The two main basins are at Sfax and El Kalaa, where they are drained by natural evaporation. However, despite the shallow depths adopted in these basins, evaporation is not satisfactory, as a thin film of oil forms on the surface of the basins, hindering the evaporation phenomenon. In Tunisia, another, more reliable solution is the direct application of wastewater from these mills to the soil.

A SADT-type analysis enabled us to model product and data flows in the three phases from global to detailed. The level of detail enables us to specify the role of each piece of data, product and element in the farming phase, the production phase and the waste management phase, taking into account the return loops of these phases to integrate the circular economy (Appendix 1).

8.4.2. *Alternatives proposed by decision-makers in the field*

Eight scenarios were extracted from the scientific literature on the basis of field studies with decision-makers (farmers, mills, the state, etc.). These scenarios have been validated by experts from the Instituts De l'Olivier De Sfax[11]. Each scenario includes a complete olive oil life cycle. Table 8.1 represents the most common situations in the local production chain, including the three main phases of the olive chain: agricultural cultivation, olive oil production and waste treatment.

11 The Institut de l'Olivier is a research institute specialized in olive growing, oleotechnics and fruit arboriculture in Sfax, Tunisia.

<table>
<tr><th rowspan="2">Phase</th><th rowspan="2">Process</th><th colspan="8">Scenarios (density between 17 and 34 Olivier /ha)</th></tr>
<tr><th>S1</th><th>S2</th><th>S3</th><th>S4</th><th>S5</th><th>S6</th><th>S7</th><th>S8</th></tr>
<tr><td rowspan="8">Farming phase</td><td>Farming practices</td><td colspan="4">Conventional</td><td colspan="4">Organic</td></tr>
<tr><td>Productivity</td><td colspan="2">900 kg/ha</td><td colspan="2">2,160 kg/ha</td><td colspan="2">1,660 kg/ha</td><td colspan="2">850 kg/ha</td></tr>
<tr><td>Irrigation</td><td colspan="2">Dryer</td><td colspan="2">With irrigation</td><td colspan="2">With irrigation</td><td colspan="2">Dryer</td></tr>
<tr><td>Soil management</td><td colspan="8">Mechanical engineering with machine operation</td></tr>
<tr><td>Size</td><td colspan="8">Manual with saws</td></tr>
<tr><td>Harvest</td><td colspan="8">Manual with workers</td></tr>
<tr><td>Fertilization</td><td colspan="4">Distribution of manure and mill wastewater with NPK and on the soil</td><td>With manure and wet pomace compost</td><td>With wet pomace compost and mill wastewater</td><td>With manure and wet pomace compost</td><td>With wet pomace compost and mill wastewater</td></tr>
<tr><td>Pesticides and herbicides</td><td colspan="4">Manual, twice a year, with insecticides (dimethoate)</td><td colspan="4">Nothing</td></tr>
<tr><td>Production phase</td><td>Processing method</td><td>Traditional system</td><td>Three-phase system</td><td>Traditional system</td><td>Three-phase system</td><td colspan="4">Two-phase centrifugation system</td></tr>
<tr><td rowspan="4">Waste management phase</td><td>Pomace</td><td colspan="4">Second pomace oil extraction</td><td colspan="4">Nothing</td></tr>
<tr><td>Wet pomace</td><td colspan="4">Nothing</td><td>Composting wet pomace with manure</td><td rowspan="2">Composting wet pomace with mill wastewater</td><td>Composting wet pomace with manure</td><td rowspan="2">Composting mill wastewater with wet pomace</td></tr>
<tr><td>Mill wastewater</td><td colspan="5">Application of mill wastewater directly on fields</td><td>Application of wastewater on fields</td></tr>
<tr><td>Treatment of pruning waste</td><td colspan="2">Crushing and scattering on fields</td><td colspan="2">Burning and scattering of ashes on fields</td><td colspan="2">Crushing and scattering on fields</td><td colspan="2">Burning and scattering of ashes on fields</td></tr>
</table>

Table 8.1. *Description of scenarios*

8.4.3. *Sustainability criteria (LCSA)*

The assessment of environmental, economic and social performance is based on the understanding and knowledge of multiple stakeholders acquired through statistics and our surveys.

8.4.3.1. *Environmental life cycle assessment*

An environmental life cycle assessment is generally used to identify environmental impacts, in order to determine sustainability assessments. Decision-makers and experts have selected the methodologies that suit them best, including the CML 2 characterization[12] and the ReCiPe 2008 methodologies[13].

Five environmental impact categories were obtained, as defined by the CML 2 baseline 2000 V2.04 characterization: ozone layer depletion, global warm, eutrophication, acidification and abiotic depletion. The results in this section were developed and analyzed using SimaPro 7.2 software[14] (Salomone and Ioppolo 2012), as shown in Figure 8.2.

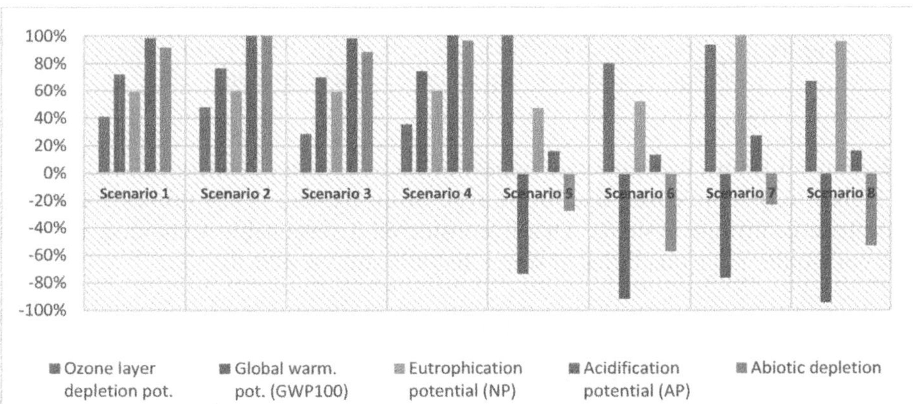

Figure 8.2. *Environmental impact of each scenario. For a color version of this figure, see www.iste.co.uk/koubaa/optimization.zip*

The eight scenarios are compared by functional unit (1,000 kg of olives). The results obtained are presented in Figure 8.2 and show, in general, that the conventional scenarios (scenarios 1–4) present the highest environmental loads for

12 This is an impact calculation method available in SimaPro.

13 This is a type of impact category available in SimaPro.

14 SimaPro is a software package for assessing the environmental impact of a process.

each impact category studied, while the organic scenarios (scenarios 5–8) present negative environmental loads.

8.4.3.2. *Life cycle cost analysis*

As with the selection of environmental criteria, decision-makers/stakeholders can choose appropriate economic criteria to assess the economic pillar, including investment costs, raw material costs, production costs and operation and maintenance costs. In this study, decision-makers and experts selected economic criteria such as operating costs (Tunisian Dinars) and productivity (kg/Ha). The results of this section were developed through surveys and interviews with stakeholders (Figure 8.3).

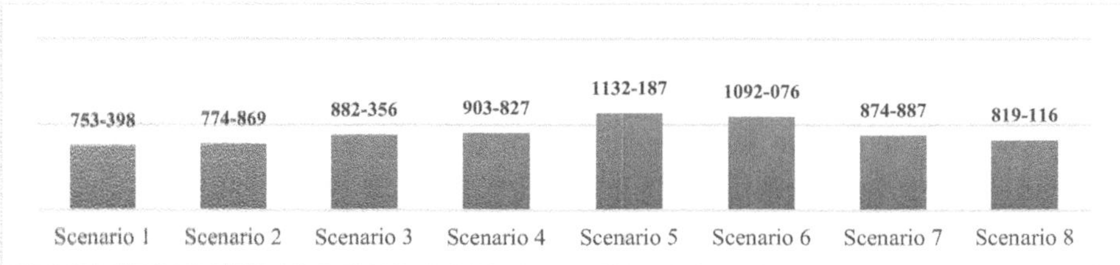

Figure 8.3. *Economic impact of each scenario15. For a color version of this figure, see www.iste.co.uk/koubaa/optimization.zip*

Figure 8.3 shows the economic impact of each scenario. Scenarios 5 and 6 have the highest economic costs, due to the type of investment in the processing phase (two-phase centrifuge system). This is considered a sustainable technique, since it does not require large quantities of water, but it is still very costly.

8.4.3.3. *Social analysis of the life cycle*

In this study, decision-makers and experts selected social criteria such as: workers (human rights), working conditions, socio-economic impacts and governance. The results of this part of the study were developed through surveys and interviews with stakeholders. In our survey, a Likert scale[16] had five points to express the degree of agreement or disagreement relative to a statement (between 1 and 5, where 1 is the lowest score and 5 the highest).

Human rights and working conditions were less respected in scenarios S3–S6, which was linked to the higher productivity in these scenarios. These scenarios require more physical effort. For socio-economic impacts and governance, scenarios

15 Data were collected from a survey of farmers in the city of Sfax.

16 A Likert scale is a psychometric tool for measuring attitudes in individuals.

S5–S8 were the best placed, since these scenarios were organic and this type of practice was highly recommended with a very high cost price (Figure 8.4).

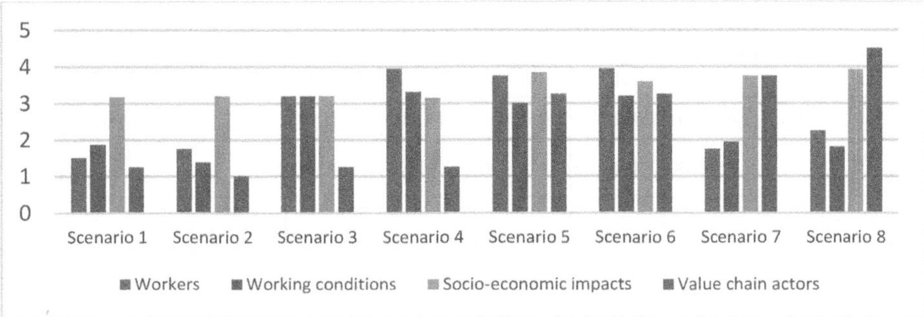

Figure 8.4. *Social impact of each scenario[17]. For a color version of this figure, see www.iste.co.uk/koubaa/optimization.zip*

8.4.4. *Results*

The BWM was used to obtain the weight of each criterion, by applying equations [8.1] to [8.4]. Table 8.2 shows the weights of the criteria in our case study.

Criteria	Weight of criteria
Environmental criteria	0.56
Economic criteria	0.33
Social criteria	0.11

Table 8.2. *Weight matrix for the three criteria*

Having obtained the decision matrix, presented in Table 8.3, the VIKOR method can be applied to evaluate the ranking of the eight scenarios in order to obtain the best compromise solution. The results of the VIKOR calculation are shown in Table 8.4. On the basis of the results presented in Table 8.4, it can be seen that scenario 3 is the best ranked according to the Q_J value (i.e. the minimum value) according to equation [8.11]. The following two conditions were examined: based on the ranking of the different suppliers, with the value $\upsilon = 0.5$, as a compromise solution. Furthermore, this alternative is the best ranked by S and R, and $Q(S4)\text{-}Q(S3) = 0.213 > 1/(8\text{-}1) = 0.143$. Therefore, satisfying both conditions 1 and 2.

17 Data were collected from a survey of farmers in the city of Sfax.

	Sub-criteria	S1	S2	S3	S4	S5	S6	S7	S8
Env	Ozone layer depletion	6.12E-05	7.16E-05	4.25E-05	5.30E-05	1.50E-04	1.20E-04	1.40E-04	1.00E-04
	Global warming	4.81E+02	5.10E+02	4.67E+02	4.96E+02	-4.92E+02	-6.14E+02	-5.10E+02	-6.32E+02
	Eutrophication	4.43E+01	4.46E+01	4.43E+01	4.46E+01	3.54E+01	3.91E+01	7.49E+01	7.19E+01
	Acidification	1.06E+01	1.08E+01	1.06E+01	1.08E+01	1.69E+00	1.42E+00	2.90E+00	1.71E+00
	Abiotic depletion	2.25E+00	2.46E+00	2.17E+00	2.37E+00	-6.80E-01	-1.41E+00	-5.70E-01	-1.31E+00
Eco	Operating cost (Tunisian Dinars)	753,398	774,869	882,356	903,827	1,132,187	1,092,076	874,887	819,116
	Productivity (kg)	900	900	2,160	2,160	1,660	1,660	850	850
Soc	Workers (human rights)[18]	1.5	1.75	3.19	3.94	3.75	3.94	1.75	2.25
	Working conditions[14]	1.875	1.375	3.19	3.31	3	3.19	1.94	1.81
	Socio-economic impacts[14]	3.17	3.2	3.2	3.15	3.84	3.6	3.75	3.92
	Governance[14]	1.25	1	1.25	1.25	3.25	3.25	3.75	4.5

Table 8.3. *The decision matrix from our survey. For a color version of this table, see www.iste.co.uk/koubaa/optimization.zip*

	S_j equation [8.9]	R_j equation [8.10]	Q_j equation [8.11]	Ranking
S1	0.538	0.159	0.791	6
S2	0.636	0.159	0.971	8
S3	0.415	0.056	0.095	1
S4	0.508	0.066	0.308	2
S5	0.474	0.165	0.702	5
S6	0.374	0.148	0.438	3
S7	0.550	0.165	0.842	7
S8	0.364	0.165	0.500	4
S+, R+	0.36	0.06		
S-, R-	0.64	0.17		

Table 8.4. *VIKOR evaluation results and sorting*

18 A five-point Likert scale.

Scenario 2 is the least sustainable life cycle, mainly because the olive oil processing method (three-phase centrifugation system) and waste recovery (second pomace oil extraction + application of mill wastewater directly to the fields + crushing and dispersion on the fields) caused both severe environmental pollution and significant economic costs. Scenario 3 is comparatively better than the other scenarios, performing particularly well in environmental, economic and social terms.

8.4.5. *Discussion*

8.4.5.1. *Sensitivity analysis for criteria weights*

A sensitivity analysis is carried out by assigning different weight vectors to the criteria, and the following four cases are examined:

– Equal importance: assign an equal weight of 0.333 to each of the three criteria (case 1).

– A dominant with the rest of equal importance (case 2, case 3 and case 4): this means that a dominant weight is assigned to one criterion and an equal weight is assigned to the three remaining criteria. In case 1 (i=2, 3, 4), a dominant weight of 0.40 is assigned to the (i-1) criterion and an equal weight of 0.30 is assigned to each of the other two criteria. Taking case 2 as an example, a dominant weight of 0.40 is assigned to environmental impacts and a weight of 0.30 is assigned to each of the other two criteria (economic and social).

According to the results presented in Table 8.5, scenario 3 is the most sustainable, and scenario 2 is the worst in all four cases. We can therefore conclude that the proposed method for calculating criteria weights is valid in terms of sensitivity.

Weight			Case	Scenarios							
Env	Eco	Soc		S1	S2	S3	S4	S5	S6	S7	S8
0.333	0.333	0.333	Case 1	2	8	1	7	5	4	6	3
0.400	0.300	0.300	Case 2	7	8	1	2	5	3	6	4
0.300	0.400	0.300	Case 3	7	8	1	2	5	4	6	3
0.300	0.300	0.400	Case 4	7	8	1	2	5	3	6	4

Table 8.5. *Results of sensitivity analysis by modifying criteria weights. For a color version of this table, see www.iste.co.uk/koubaa/optimization.zip*

8.4.5.2. *Robustness analysis*

In this study, the TOPSIS method was used to verify the performance of the VIKOR analysis in reinforcing the three environmental, economic and social objectives. The coefficients calculated for the eight scenarios were (S1=0.348, S2=0.322, S3=0.6678, S4=0.648, S5=0.482, S6=0.640, S7=0.334 and S8=0.620). From the results of the full evaluation model, the sequence of olive oil life cycles, from best to worst, was derived as follows: scenario 3 (best), scenario 4, scenario 6, scenario 8, scenario 5, scenario 1, scenario 7, scenario 2 (worst). Subsequently, the life cycle (scenario 3) was obtained as the best life cycle that could be recommended for olive oil production in Tunisia. The ranking order was the same as for VIKOR (S3, S4, S6, S8, S5, S1, S7 and S2). We can therefore conclude that the proposed method for ranking alternatives is valid in terms of robustness.

8.5. Conclusion

This study uses an MCDM methodology for LCSA of olive oil production in Tunisia. MCDM and LCSA are integrated to assess the sustainability of agricultural practices, processing methods and waste recovery management, so that all life cycle stages of the olive oil chain are circular and sustainable. The proposed method enables decision-makers/actors to identify the most sustainable scenarios for achieving their objectives from among different alternatives.

The LCSA tool combines E-LCA, LCCA and S-LCA. The study addresses the three pillars of sustainability – environmental aspects, economic issues and social concerns – through E-LCA, LCCA and S-LCA, respectively. The criteria of the three pillars can be obtained from a life cycle perspective as elements of the decision-making matrix.

The BWM and VIKOR methods are combined to determine the sustainable sequence and the most sustainable scenario. BWM is used to calculate the criteria weights for the sustainability assessment. VIKOR is used as the MCDM method to rank alternatives in descending order, according to the decision matrix determined by LCSA.

We also applied a sensitivity analysis to check against variations in the weights of the evaluation criteria. This sensitivity analysis of the weights revealed that even significant changes in the weights did not produce a big difference in the ranking given by the proposed methodology. The next step is to identify interactions between decision-makers and between criteria. An extension of linear programming in alignment with the VIKOR method for calculating weights, and by using another method like BWM, we can reduce the subjectivity of multi-criteria methods.

8.6. Appendix

8.6.1. *Olive oil life cycle*

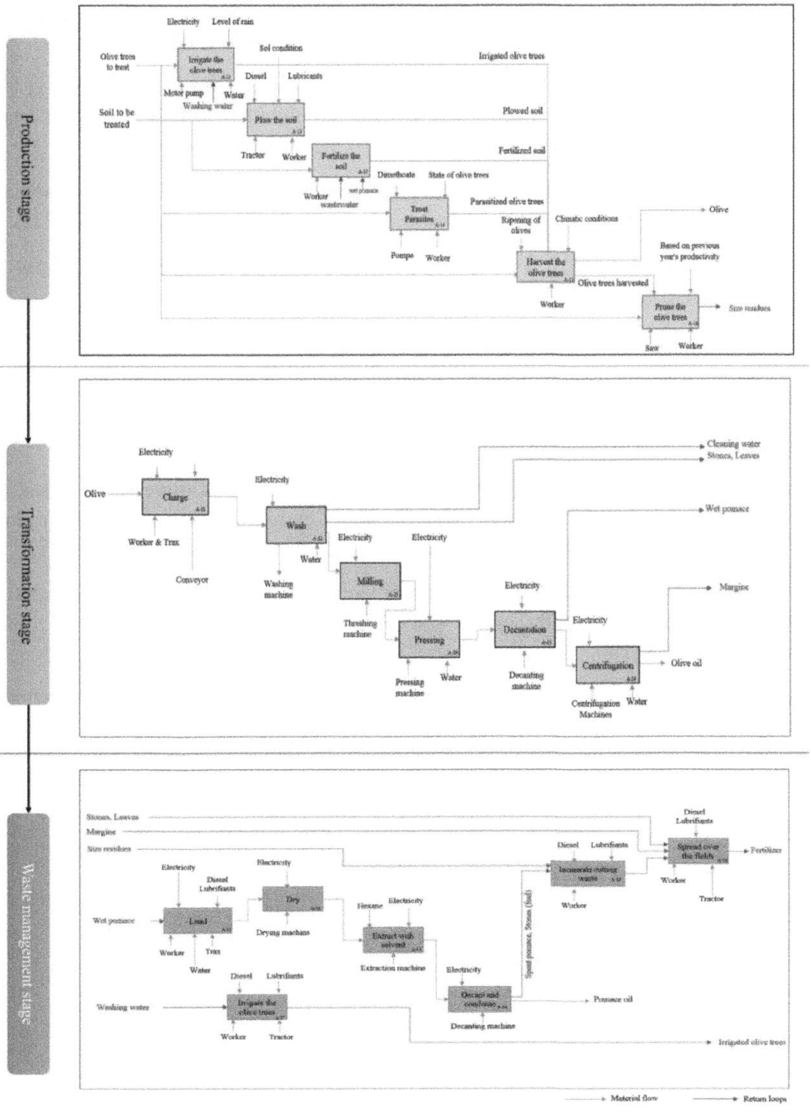

Figure 8.5. *Olive oil life cycle. For a color version of this table, see www.iste.co.uk/koubaa/optimization.zip*

8.7. References

Duman, K., Özgen, Ö., Üçtuğ, G. (2020). Environmental life cycle assessment of olive pomace utilization in Turkey. *Sustainable Production and Consumption*, 22, 126–137.

Fernández-Tirado, F., Parra-López, C., Romero-Gámez, M. (2021). A multi-criteria sustainability assessment for biodiesel alternatives in Spain: Life cycle assessment normalization and weighting. *Renewable Energy*, 164, 1195–1203.

Foolmaun, K. and Ramjeawon, T. (2013). Life cycle sustainability assessments (LCSA) of four disposal scenarios for used polyethylene terephthalate (PET) bottles in Mauritius. *Environment, Development and Sustainability*, 15(3), 783–806.

Jellali, A., Hachicha, W., Aljuaid, M. (2021). Sustainable configuration of the Tunisian olive oil supply chain using a fuzzy TOPSIS-based approach. *Sustainability*, 13(2), 722.

Rapa, M. and Ciano, S. (2022). A review on life cycle assessment of the olive oil production. *Sustainability*, 14(2), 654.

Ren, J. and Toniolo, S. (2018). Life cycle sustainability decision-support framework for ranking of hydrogen production pathways under uncertainties: An interval multi-criteria decision making approach. *Journal of Cleaner Production*, 175, 222–236.

Rezaei, J. (2015). Best-worst multi-criteria decision-making method. *Omega*, 53, 49–57.

Riesgo, L. and Gallego-Ayala, J. (2015). Multicriteria analysis of olive farms sustainability: An application of TOPSIS models. *Handbook of Operations Research in Agriculture and the Agri-Food Industry*, Plà-Aragonés, L.M. (ed.). Springer, New York, 327–353.

Salomone, R. and Ioppolo, G. (2012). Environmental impacts of olive oil production: A life cycle assessment case study in the province of Messina (Sicily). *Journal of Cleaner Prodution*, 28, 88–100.

Zhang, H. and Haapala, R. (2015). Integrating sustainable manufacturing assessment into decision making for a production work cell. *Journal of Cleaner Production*, 105, 52–63.

9

Open Innovation as a Solution in the Food Supply Chain

In a rapidly changing economic environment, food supply chains face numerous challenges due to globalization, lower trade barriers and various economic, health and political crises. This forces businesses to adopt specific management practices that promote the effectiveness of innovation and thus sustainable growth. All of these factors are driving companies to open up their innovation processes to collaborate with various external partners such as customers, suppliers, universities and stakeholders. The importance of the topic and its motivation arise, on the one hand, from the need to promote the economic development of the agri-food sector and, on the other hand, from the majesty of open innovation to ensure the further development of the sector.

9.1. Introduction

In an increasingly changing and complex global context, players in socio-economic markets have developed partnership strategies that are more focused on open innovation processes to share risks and maximize profits. This strategy is underpinned by a new international literature on open innovation, which promotes collaboration between different actors outside the company as an essential element for generating (Bogers et al. 2018; Pustovrh et al. 2020). The recent crisis triggered by the Covid-19 pandemic has shown that companies must increasingly build capacity and resilience to maintain their competitive advantage and survive in a dynamic and changing context (Labrouche 2016; Bertello et al. 2022). In addition, new structures (startups and SMEs) face major challenges and opportunities, which require paradigm shifts, new states of thinking and new practices to be leveraged

---

Chapter written by Bouthaina MAAZOUN and Raoudha KAMMOUN.

from open knowledge and scientific channels through the co-creation of external technologies, solutions and innovations. Therefore, the co-creation of external technologies, solutions and innovations generate open knowledge and scientific channels. Innovation is increasingly becoming a key factor to maintain customer satisfaction, reduce costs and improve long-term competitiveness and performance (Porter 1985; Dosi et al. 2000; Howells 2005).

It should be pointed out that the innovation concept, originally focused on closed innovation, emphasizing the linear innovation process with large investment in the R&D process, has evolved into open innovation. The new concept creates varying degrees of organizational sensitivity, increasing or decreasing the flow of innovation inputs or products (Jamett et al. 2017). It describes how a company interacts with external organizations to achieve its innovation goals with fewer resources and with new market outcomes (Chesbrough 2007). In the field of innovation management, open innovation is becoming increasingly important (Bogers et al. 2017). In fact, it affects multiple levels and covers different aspects, which complicates its integration (Randhawa et al. 2016).

For numerous years, the open innovation model has attracted great interest from the scientific and industrial communities (Liu and Vrontis 2017). Despite the importance of the food chain and the entire food sector, the number of case studies of food companies involving this innovative approach is still small compared to other fields (Sarkar and Costa 2008; Bresciani 2017; Misra and Mention 2021).

In the agri-food sector, long considered a sector with low research intensity (Minarelli et al. 2015; Bigliardi et al. 2020) and low technological intensity compared to other sectors (Cohen and Klepper 1996; Misra and Mention 2021), innovation is gaining more and more importance and considerable interest. It should be noted that the agri-food sector is influenced by a variety of factors, such as the growth of the world population (Grainger 2010), intense economic competition (Läpple et al. 2015), the limitation of fossil resources (Preschitschek et al. 2011), climate change and its possible effects on food security (Knickel et al. 2009). In this context, it is necessary to increase the production of food, fibers and energy by optimizing the consumption of available resources. To do so, companies in the food supply chain are required to increasingly promote innovation throughout their processes (Roucan-Kane et al. 2013). Furthermore, some authors (e.g. Meynard 2017; Faure et al. 2018a, 2018b; Saguy 2022) argue that the specificities characterizing the food supply chain today increasingly require the promotion of collaborative innovation. Recent studies highlight the ability of the food and beverage sector to take advantage of knowledge based on information technology to develop open innovation strategies (Cillo et al. 2019). In this perspective, various innovation initiatives have been developed, more and more frequently, in this sector despite the fact that it is considered a segment with low technological development. These initiatives involve various actors, such as customers, suppliers and universities (Johnston 2020; Saguy 2022). The current socio-economic context and market disruptions highlight the needs of food chain actors for a new mindset to adopt open innovation and new strategies of collaboration to co-create a very prosperous future (Saguy 2022).

Various research works have examined the effect of open innovation practices in companies (Fortuin and Omta 2009; Bigliardi et al. 2012; Bigliardi and Filippelli 2022a). These research works show that open innovation is the appropriate paradigm to address the new challenges facing the agri-food sector (Bayona et al. 2017; Bresciani 2017a). It should be noted that agri-food companies have particular specificities which make them more vulnerable in the context of a global, changing and unstable environment, characterized by the dependence of the production process on an incontrollable factors such as the perishability of products and raw materials (Grunert et al. 2008; Bigliardi and Galati 2016)). Moreover, these companies operate in a more turbulent market, characterized by global competition, rapid consumption, high sustainability requirements in relation to consumer health, animal welfare and the environment (Miglietta et al. 2017).

Although the agri-food sector is booming in these difficult times, food producers, due to the high competition in the market, are looking for ever-more effective and efficient management practices (Gianni et al. 2017b). Therefore, in order to improve the quality and ensure food safety, an important criterion on a global scale (Salampasis et al. 2015), the adoption of management standards extends to all companies in the agribusiness sector (Gianni et al. 2017a). Indeed, the food chain is governed by strong interdependencies between its levels, with retailers and control bodies imposing various standards and regulations (Gianni et al. 2017a). In this context, Kafel (2013) stipulates that companies in the agri-food sector implementing management systems reach a higher level of maturity and offer increased financial performance.

The agri-food sector is considered to be a vital sector in Tunisia. This sector is widely considered as a vector of the economic development of the country. Tunisia's openness to international trade and its integration into global value chains have made strong progress since the mid-1990s, reflecting the country's comparative advantages (Lopes et al. 2017). Exports have increased significantly, with a move up market and diversification, driven by the manufacturing sector, thus this reveals high growth potential (Lopes et al. 2017).

9.2. Theoretical and practical approaches of innovation

According to the neoclassical theory, localization (the territorial component) and endogenous potential have a significant impact on innovation and can be analyzed and understood multi- and cross-disciplinarity.

Management science literature shows that the construction of innovation is an important concern of researchers and practitioners in organizational strategy and management. Several authors have defined innovation in different senses, from Schumpeter (1934) who considered innovation to be absolute and dependent on certain internal and external variables, to Millier (2004) who placed innovation at the heart of all entrepreneurial activity, then to Covin and Slevin (1991) in their work on entrepreneurial orientation, linking it to the ability to bring a new product or service to market.

In terms of development theory, Schumpeter is renowned for his contributions. Specifically, the concept of innovation is a major focus of his work. Schumpeter identified two key elements of innovation: the uncovering of new scientific and technical knowledge, which he calls "invention", and the development and utilization of this knowledge, termed "innovation".

According to the New Oxford Dictionary of English[1], innovation is the process of bringing novelty to alter products or processes. It can be a major or incremental change made by big businesses, independent contractors or private individuals. In addition, innovation can be relevant for both the public and private sectors (hospitals, social centers, town halls, etc.). It can be radical (does everything) or incremental (does something better than nothing) and can occur at a variety of levels (organizations, management groups and departments, regions, localities, individuals, etc.).

The OECD proposes another approach to innovative concepts as follows: it may represent the implementation of new or significantly improved products, new processes, etc. A new approach to marketing or a new way of organizing business practices in an organization workplace or external relations (OECD 2013).

Innovation is the achievement of results through the process of change. The added value of a product is realized when the concept of reality is transformed into a tangible form. In the management of innovation, creative destruction has the same meaning as innovation. Creating and promoting is the number one priority of every business. A new concept to replace the outdated one is thus needed. New ideas are often the best and can lead to success. A dynamic and evolutionary perspective

1 New Oxford Dictionary of English (1998), p. 942.

surrounds current innovative concepts. In order to compete, companies need to embrace this mentality. Consistently replacing and innovating old technologies, products and processes because it is critical. Innovation can be effectively achieved through continuous learning and understanding of its implications.

9.3. From closed innovation to open innovation

This section aims to highlight the context, the typology and the challenges of open innovation.

9.3.1. *Context of open innovation*

The concept of closed innovation has worked for a long time, assuming that the company is able to master its entire environment. This closed strategy has proven difficult to implement in complex environments characterized by continuous technological changes, with increasingly efficient information and communication technologies (Quiñones et al. 2019). Indeed, an innovative project cannot be the result of a single organization that does all the work. On the contrary, it is the result of the interaction between a wide range of actors who cooperate in different forms, more or less formal (licenses, formal and informal collaborations, etc.).

The idea of exploiting external sources of innovation, i.e. resources located outside the company's borders (inside-out) is not new (Jamett et al. 2017). A simple historical reading of the development of technical systems makes it possible to affirm that openness is an intrinsic dimension of innovation. Historians of science and technology (Lawrence et al. 1984; Dumas 2002) have clearly highlighted the collective character associated with the emergence of technical systems and the resulting clusters of innovations. Since the creation of the cluster concept in the 1990s, along with the emphasis on globalization, organizations have been able to focus their innovation management on a logic of collaboration and the generation of agreements, so that the emergence of new methods of generating innovation accelerates its evolution.

In this sense, Cohen and Levinthal (1990) were interested in studying the absorptive capacity of companies, which means their ability to recognize the value of information obtained from outside by exploiting it within the company. In addition, March (1991) was interested in studying the effect of the exploration of new possibilities such as the exploitation of old notions related to organizational learning. Teece et al. (1997), on the other hand, explored other aspects of innovation, such as dynamic capabilities, meaning that firms have the ability to define internal and external skills and integrate them into a changing and evolving environment.

Open innovation appears as a way for organizations to extend their limits and their possibilities to innovate (Chesbrough and Appleyard 2007). This proposal is based on the new competitive environment of value creation that forces companies to seek new opportunities to organize internal processes and design their relationships with other agents, with whom they will share risks and resources (Minshall et al. 2007). According to Ollila and Yström (2016), this new global approach to innovation consists of "combining knowledge, technologies and other resources across organizational boundaries". The fundamental principle is that knowledge and useful information are widely spread. The idea of unconsciously seeking external resources to implement internal processes, as well as marketing internal opportunities is more recent and marks the time when the concept of "open innovation" was born (Huizingh 2011; Zhao et al. 2015).

It should be noted that some companies are faced with a dilemma when they find themselves forced to choose between a closed and controlled R&D environment or, on the contrary, opening their innovation process (Almirall and Masanell 2010). The transition from closed innovation to open innovation requires the participation of professionals in decision-making related to innovation activities (Huizingh 2011; Ahn et al. 2017). It involves certain perspectives, such as: the globalization of innovation, the outsourcing of R&D, the involvement of suppliers and the user as a source of innovation and the commercialization and application of external technologies (Gassmann 2006).

Open innovation models are classified by Martínez and Guadalupe (2010) into two groups, those "strategic" where cooperation leads to agreements or alliances defined almost always in a contractual document and those "organizational" which promote receptivity. At the beginning, the first models observed in practice are characterized by the highlighting of networks and connected groups that potentiate a common search around innovation, and later they became more organized resulting in models with a structure that provides a concrete framework for innovation activities carried out together, becoming an integral part of the organization of each of the participating entities or subject to independent organization (Jamett et al. 2017). The comparison between closed innovation and open innovation is shown in Figure 9.1.

It should be noted that the concept of open innovation was introduced by Chesbrough in 2003, who emphasized the importance and impact of external factors of innovation. This is because actionable ideas can come from and can also be marketed inside or outside the organization. This is a new paradigm which is imperative for companies that want to innovate successfully. The literature review reveals several definitions of open innovation (Dahlander and Gann 2010) that have evolved over the years (Tynnhammar 2019). Table 9.1 summarizes the evolution of these definitions.

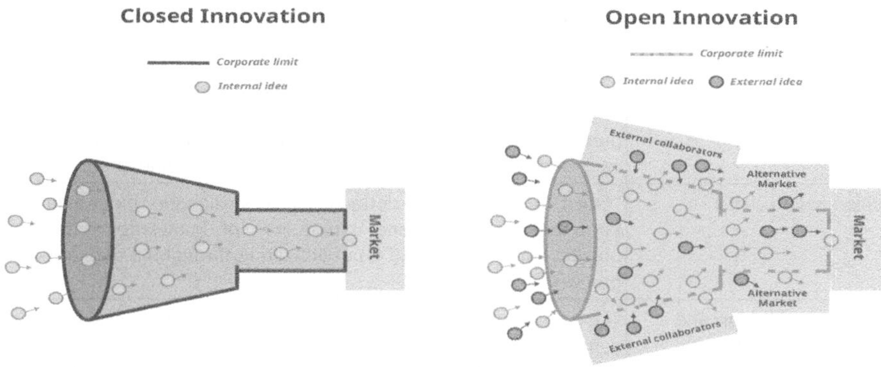

Figure 9.1. *Comparison between closed innovation and open innovation (source: https://www.viima.com/blog/open-innovation). For a color version of this figure, see www.iste.co.uk/koubaa/optimization.zip*

Although these definitions seem different, they are all centered on the fact that innovation becomes open since it becomes less and less carried out internally and becomes part of a collective approach involving many actors including competitors, public research laboratories, suppliers, in the different stages of the process of its development.

It should be noted that the field of open innovation has expanded considerably, involving a range of innovation practices and expanding the scope of analysis at all levels of the company, from the individual level to the societal level (Bogers et al. 2018). Links have been made with open source (West and Gallagher 2006), distributed user innovation (Hippel 2005), open strategy and open science (Chesbrough 2017).

Over the years, the phenomenon of open innovation has attracted more and more attention in innovation management (Popa et al. 2017b). It is a rapidly developing area of research (Bogers et al. 2017), which can be evidenced by the growing number of academic publications and special issues in journals (Cheng and Huizingh 2014).

Reference	Definition of open innovation
Chesbrough (2003)	"Open innovation is a paradigm that assumes that companies seeking to advance their technology can and should use external ideas as well as internal ideas and internal and external pathways to market."
Chesbrough et al. (2006)	"Open innovation is the use of intentional inflows and outflows of knowledge to accelerate internal innovation and expand markets for external use of innovation, respectively."
Chesbrough (2007)	"Open innovation is the pooling of knowledge for innovative purposes, where contributors have access to each other's inputs and cannot exercise exclusive rights to the resulting innovation."
Lichtenthaler (2017)	"Open innovation is defined as a systematic exploitation of knowledge, retention and exploitation of knowledge within and outside the boundaries of an organization throughout the innovation process."
San-Martín-Albizuri and Rodríguez-Castellanos (2012)	"Open innovation is the set of actions and practices by which companies incorporate agents of innovation processes who have not traditionally participated in them, whether internal or external to the company."
Deutsch (2013)	"Open innovation is the discipline and practice of leveraging the non-obvious results of others as the first step in an innovation process through formal and informal relationships within an organization."
Chesbrough and Bogers (2014a, 2014b)	"Open innovation is a distributed innovation process based on purposefully managed knowledge flows across organizational boundaries, using pecuniary and non-pecuniary mechanisms consistent with the organization's business model."
Tynnhammar (2017)	"Open innovation is the sharing of knowledge, across organizational boundaries, within a business model involving many partners working together throughout the development process."

Table 9.1. *Evolution of definitions of open innovation (sources: Tynhammar (2017); Lopes and de Carvalho (2018))*

9.3.2. *Typology of open innovation*

Different researchers have emphasized the diversity of forms of open innovation. For example, Gassmann and Enkel (2004) emphasized a classification according to the strategy adopted. To do this, they distinguish between the incoming strategy (incoming open innovation), the outgoing strategy (outgoing open innovation) and the coupled or mixed strategy (mixed open innovation).

– **Incoming open innovations** are related to the search for useful external information (West and Bogers 2013) and can include the purchase of patents and licenses (Jamett et al. 2017), supplier innovation competitions (Bullinger-Hoffmann et al. 2009), co-creation with customers or crowdsourcing (Christensen and Karlsson 2019), through which a company can outsource the creative process to a multitude of website users in the form of an open call on the Internet (Buecheler et al. 2010).

– **Outbound open innovations** are linked to activities such as spillovers from the sale of patents or licenses or joint-venture activities (Chesbrough and Brunswicker 2014). These activities are carried out with the aim of maximizing the potential of innovation activities for the competitive advantage of innovative firms (Kutvonen 2011).

– **Coupled (or mixed) open innovations** consist of coupling incoming and outgoing open innovation (Wikhamn and Wikhamn 2013) with co-creation partnerships (Lee et al. 2012). In addition, new open innovation tools and mechanisms have emerged, such as innovation intermediaries (Laursen and Salter 2006), toolkits (Piller and Walcher 2006), patent actions (Arora and Gambardella 2010) and open platforms (Gawer and Cusumano 2014), which facilitate and generate value creation, capture and distribution across the borders of a single organization.

Another classification of open innovation is proposed by Vrande et al. (2009). It highlights a dichotomy that opposes exploitation and exploration:

– **Exploitation** includes, for example, the use of spin-off companies and the licensing of intellectual property are commercially oriented (Long et al. 2020). Its incoming nature provides information and improves knowledge, which could facilitate responsiveness (Jamett et al. 2017).

– **Exploration** includes customer involvement in R&D, external networking, outsourcing of R&D and licensing of intellectual property. It brings information and knowledge to the innovative company, provides contacts with stakeholders and thus offers the possibility of increased anticipation (Gupta et al. 2016). In addition, open innovation exploration activities involve attracting collaborators to expand and use the innovative organization's network and build relationships with stakeholders.

In practice, there are different ways to implement open innovation, which are used arbitrarily by organizations and where applications are sometimes intuitive (Jamett et al. 2017). Furthermore, Jamett et al. (2017) stipulated that the open innovation initiative requires strong management commitment and support, restructuring and redefining of roles and formal strategies materialized by clear objectives. This requires us to take into account the stages of change that generate a cycle capable of breaking inertia and maintaining systematization mechanisms over time (Jamett et al. 2017). It should be noted that in order to benefit from the advantages provided by the acquisition of external knowledge from open innovation, the company is required to have the internal expertise necessary to use this knowledge effectively (Lopes and

De Carvalho 2018). Having the ability to combine internal and external sources of information is a specificity that can generate a competitive advantage for the company (Cassiman and Veugelers 2006; Lestari et al. 2020).

9.3.3. *Challenges of open innovation*

The literature review reveals different issues with open innovation. Vrande et al. (2009) were interested in exploring the problems related to open innovation practices such as lack of time, administrative burdens and lack of resources, as well as those related to culture and organization within companies. Their research consists of an exploratory study with a large sample of Dutch companies. These authors found that the problems recurring in all of the companies studied are mainly related to the management of time and resources. In their research, which consists of a case study, Dufour and Son (2015) were interested in providing information on the obstacles related to open innovation and the ways to overcome them by analyzing the various organizational changes undertaken by the company. These authors previously identified four potential barriers to open innovation in the literature: organizational structure, knowledge management systems, corporate culture and networking. In 2016, Bigliardi and Galati carried out a survey of 157 Italian SMEs aimed at carrying out a typological analysis in relation to the obstacles linked to the adoption of open innovation. This cluster is made up of traditional and less innovative small- and medium-sized enterprises. The authors also identified four main categories of challenges: knowledge, collaboration, organizational and financial strategy. They concluded that knowledge-intensive industries consider the lack of financial resources, the strategy related to the implementation of open innovation and the lack of knowledge to be the main barriers for the adoption of open innovation. Bertello et al. (2022) longitudinally analyzed three pre-competitive projects, exploring the main challenges faced by traditional small- and medium-sized Italian enterprises, involved in university-industry collaboration in the field of basic and experimental research. In their studies, the researchers were interested in studying the impact of this collaboration throughout the four phases of the project, which are the initiation phase, the planning phase, the execution phase and the closing phase, as well as the transversal phase of monitoring and control. In the first phase, they identified a lack of innovation strategy, a lack of partner mapping for the companies studied, a non-convergence of objectives, as well as a lack of knowledge of the partners at the project level. In the implementation phase, they also identified unknown partners, inadequate information systems, underestimation of partners' efforts, and excessive diversification and heterogeneity at the project level. Finally, in the closing phase, they identified the lack of post-project commitment at the level of the companies studied, delays in the achievement of objectives and a lack of effectiveness of policies related to the high level of technological readiness of the project. Su et al. (2022) classified the challenges of open innovation in three levels:

organizational, operational and individual level. On the first level, the limitation of a lack of clear vision can cause limitations in defining and communication open innovation goals and objectives with internal and external stakeholders. The authors identified the other limitations caused by resource constraints, lack of structure, implementation challenges and compliance concerns. On the second level, the lack of community engagement may provoke limitations in participation of external actor's. The other challenges related to the second level are: coordination constraints, quality assurance, time constraints, technological affordance, identification limitations and security concerns. The main risks caused by these challenges were identified in this study. Finally, the challenges for the individual level are idea competition effect, knowledge barriers and individual risk factors. The risks caused by these challenges are limitation in the motivation of individual actors, limitation in recruiting and skills and mitigation in intellectual property rights. Figure 9.2 shows the challenges of open innovation classified according to the three levels proposed by authors.

Figure 9.2. *Open innovation challenges (source: authors inspired from Su et al. (2022)). For a color version of this figure, see www.iste.co.uk/koubaa/optimization.zip*

9.3.4. *Practices of open innovation*

The most common conventional partner identification practices are expert opinion, human relations, email inquiries and online communities (West and Borgers 2013). In addition, other more structured and contractual practices are increasingly used such as strategic alliances, joint ventures, licenses, subcontracting and others (Schilling 2007; Lopes et al. 2017). Some of these practices are developed as follows:

– **Strategic alliances** are voluntary cooperation agreements between companies (Parkhe 1993), which can be used to access knowledge or to fully exploit our own knowledge or capacities by using them in the process development (Schilling 2007). Furthermore, a company can collaborate and pool resources with another company to collectively develop a product, share risks or accelerate time to market (Schilling 2007).

– **Joint ventures** are a contractual arrangement that creates a separate legal entity in which the parent companies hold shares under the terms and conditions specified in a legal document (Carland et al. 1998). The partners who make up the pooling arrangement provide means to develop a joint business managed on an equal footing in order to generate profits.

– **Licensing** defines an agreement by which a company obtains the rights to use technology from another company and allows an organization to quickly acquire technology that it does not own (Schilling 2007). If two parties are mutually interested in knowledge or technology, a cross-licensing agreement could be entered into, with both parties granting each other licenses (Bogers et al. 2018).

– **Outsourcing** enables firms that do not have the skills or facilities to perform all activities in the value chain to develop new innovations (Schilling 2007) through access to ideas and technologies from external sources according to specifically established contracts (Jamett et al. 2017). It should be noted that subcontracting is a common form of outsourcing that allows companies to manufacture products on a large scale without making large investments, while allowing the company to specialize. Gassmann et al. (2010) argue that outsourcing is often a starting point in the open innovation process.

– **Co-creation** defines "any act of collective creativity, namely creativity shared by several people" (Sanders and Stappers 2008). However, the concept is used today to define collaboration with end-users of products or services. In fact, over the past few years, designers have moved closer to end customers during the development process (Sanders and Stappers 2008) and consumers are increasingly looking for ways "to exert influence in every part of the system commercial".

– **Networking** is a form of intermediate organization between the market and the company, characterized by the plurality of cooperation agreements between the different groups or stakeholders (suppliers, customers, competitors, public and private institutions, etc.) (Briones et al. 2011).

– **Customer engagement** refers to the actions and interactions where a development team engages with current (or potential) customers in a program, project and/or stage of the development process to seek specific information as latent needs, to develop customer knowledge and develop new solutions to the problems encountered (Sandel 2007).

– **Spin-off** is an agreement created for the purpose of commercializing one or more research discoveries outside the core business of the original company (Garvin 1983). As not all spin-off initiatives are equal, different authors have tried to categorize and establish different models and types: according to their origin (academic and industrial), the technology used (technology or conventional base), the strategy (reactive, proactive), the reasons (restructuring while safeguarding the know-how) and the impact on the strategy of the original organization (technical and competitive). Ultimately, Spencer (2008) believes that managerial capabilities are an important condition for the success of technology spillovers.

9.4. The food supply chain: characteristics and challenges

This section aims to highlight the food supply chain characteristics and challenges.

9.4.1. *Characteristics of the food supply chain*

The agri-food sector has long been considered relatively traditional, mature and characterized by a great diversity of companies and products (Bigliardi and Galati 2018) and low technological intensity compared to other sectors (Cohen and Klepper 1996). Figure 9.3 presents a schematic representation of an agri-food supply chain.

Agri-food companies are quite conservative in terms of the innovations developed (Martinez and Briz 2000; Yadav et al. 2022), however, they require the implementation of advanced technologies and the introduction of new products on the market (Wijnands et al. 2008). As a result, innovation is increasingly becoming a major concern in this sector. The anchoring of the culture of innovation within this sector is dictated by the turbulence of the environment and the socio-economic and cultural context in which agri-food companies operate. In the agri-food sector, there are several interrelated and interdependent issues which are in most cases uncontrollable, such as the perishability of products and inputs as well as the increase in global competition and the level of consumption which has become increasingly fast. To all of these factors is added the increase in legal and regulatory requirements and requirements related to sustainability (Bilali 2018, Yadav et al. 2022). The global agri-food industry operates in an environment where policies, standards, regulations, guidelines, training and guidance related to food, including those related to food safety, are being developed or updated (King et al. 2017). The need to reduce complexity and improve the efficiency and effectiveness of food chains is becoming increasingly important. Sustainable nutrition and food security are at the center of concerns and key themes of sustainable development given that around nine billion people on Earth are expected to be fed by 2050, 56% of them through sustainable and safe food chains (Godfray et al. 2010).

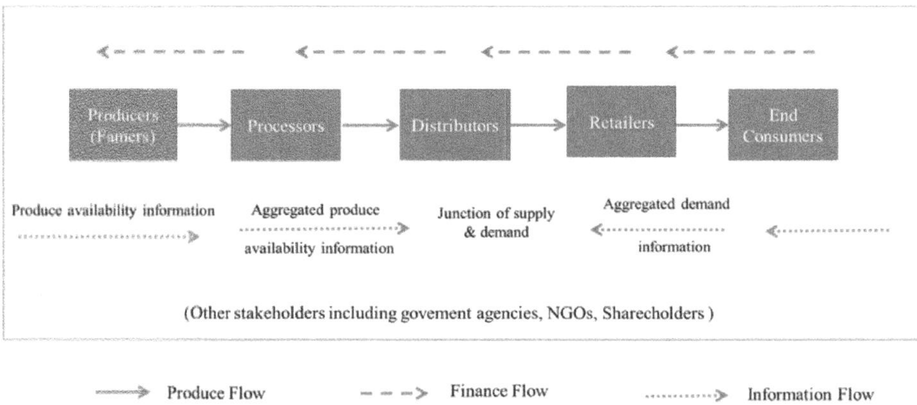

Figure 9.3. *A schematic of agri-food supply chain (source: Yadav et al. (2022)). For a color version of this figure, see www.iste.co.uk/koubaa/optimization.zip*

In order to meet the growing demand for food products associated with global population growth, industrial and centralized production systems, including large-scale agriculture, intensified animal production, large-scale processing and distribution, have developed significantly over the past few decades (Tsimiklis and Makatsoris 2016). Moreover, it is widely recognized that food production systems and the food chain in general have an obligation to become fully sustainable (Godfray et al. 2010). Food producers are then required to cope with the increasing complexity of food supply chains, by revising food safety regulations (Zach et al. 2012). Therefore, among policymakers, academics and other stakeholders, there is a consensus that food systems need to be more inclusive, practical, nourishing and beneficial for a sustainable future (Bertello et al. 2022; Misra and Mention 2021; Kumar et al. 2022).

Although the agri-food sector is booming in these difficult times, food producers, due to the high competition in the market, are looking for ever more effective and efficient management practices (Gianni et al. 2017b; Bigliardi and Filippelli 2022b). To this end, "quality" and "safety" are the main characteristics of food products that largely determine the "competitiveness" and "profitability" of the food business (Mueller and Moedas 2017). Management standards are seen as a means of expressing and disseminating recommended principles and practices. The combined adoption of quality and food safety-oriented management standards extends to a majority of organizations in the agri-food sector (Gianni et al. 2017a). These standards significantly contribute to guaranteeing the quality of food products, reaching a high level of maturity and offering increased financial and operational performance (Kafel and Sikora 2014).

At this level, it should be remembered that the agri-food chain is governed by strong interdependencies between the various stakeholders, with retailers and control bodies imposing various regulations. However, retailers act as the main drivers for the adoption of standards and the integration of management systems in food companies (Kafel 2016; Gianni et al. 2017a). Su et al. (2022) explored the major stakeholders involved in the agri-food supply chain (Figure 9.4).

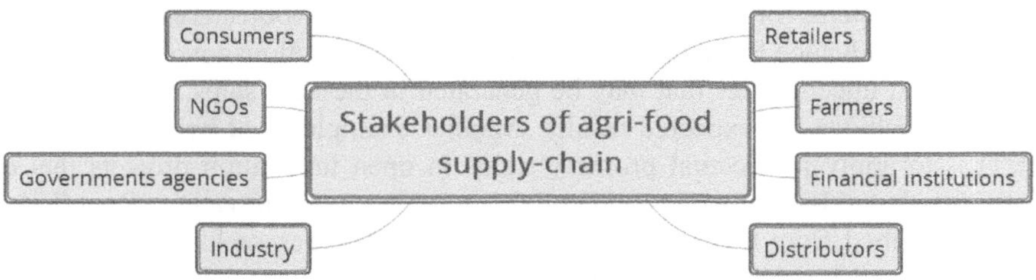

Figure 9.4. *Stakeholders of agri-food supply chain (source: authors inspired from Su et al. (2022))*

In such context, innovation enables agribusinesses to adapt to complex and structural changes in the sector, comply with new regulations and anticipate changes in consumer preferences (Sarkar and Costa 2008; Saguy and Sirotinskaya 2016). It also improves the competitiveness of these companies in order to challenge their main national and international competitors (Sánchez et al. 2016; Bayona et al. 2017).

9.4.2. *Challenges of the food supply chain*

The food supply chain is characterized by a regulated interaction obeying specific and rigorous rules. Therefore, each innovation approach undertaken should take into account these rules (Acha 2008). However, the value chain of the agri-food industry involves human and material resources, technologies and synergistic modes of organization that collaborate in the development, production and/or processing as well as the marketing of specific food products. Therefore, the major challenge is to open up innovation in this sector (Chesbrough 2003; Gupta et al. 2016). In this context, it is important to take into account emerging opportunities (shorter development time, shorter time to market, lower research and development cost and a better understanding of customer needs) and adopt them in open innovation approaches (Costa et al. 2018). These authors add that because of the novelty of this approach, it is necessary to take into consideration how the management processes the problems arising from its implementation in order to take advantage of this new paradigm. Several studies have focused on studying the challenges of open innovation (Chin and Karlsson 2020; Saguy 2022). These studies reveal recurring and interrelated problems such as the lack of support during technology transfer, the lack of financial resources as well as the management of the legal framework, especially with regard to the protection of intellectual property related to innovations and communication barriers. All of these challenges have a considerable impact on the progress of open innovation projects and can lead to a very great waste of time and/or money which can result in the failure of the initiatives undertaken. As an example, Specht et al. (2016) studied open innovation in urban agricultural systems in Berlin and found that if project objectives are correctly established in advance from the start of projects, the associations involved will avoid problems and uncertainties that may be generated in the next steps, which makes it possible to achieve the expected results. Another example cited by Montufar et al. (2021) is to apply intellectual property rights in open innovation projects that are jointly developed by companies and collaborators in open innovation. Carroll et al. (2017) described from their research the case of companies seeking to manipulate intellectual property rights to the detriment of other employees, which could lead to an increase in the vulnerability of these organizations to the risks associated with open innovation, which will have a negative impact on the long-term sustainability of these organizations. The management of intellectual property by companies ranges from confidentiality agreements to patents. Faure et al. (2018) propose another strategic scheme to reduce the risks associated with open innovation which could consist of restricting the external sources available to companies so that they can focus on the most valuable sources according to the knowledge they need to access, also providing the necessary means to acquire and absorb this knowledge. The majority of research in open innovation focuses essentially on large structures and innovative companies active in industries with high research and development intensity. It is only recently that this stream of literature has begun to open up to small- and medium-sized enterprises to better understand how open innovation approaches are integrated and implemented in smaller and more traditional companies (Dooley and O'Sullivan 2018; Marco et al. 2020). Vrande et al. (2009) analyzed the obstacles linked to open innovation for a sample of 605 Dutch small-

and medium-sized enterprises adopting open innovation. They found that organizational and corporate culture issues emerge when two or more companies work together. From which they concluded that these problems are clearly the major obstacles to open innovation, followed by the provision of time and resources and finally administrative burdens. All of these obstacles relate to specific typologies of open innovation.

Organizational and cultural barriers are manifested more clearly within companies following their participation in other companies or the involvement of external parties and users in these companies. According to the same authors, administrative burdens are closely related to the company, to its participation in other companies and to the involvement of external parties, more specifically when cooperating with governmental institutions or other institutions for nonprofit purposes or when companies receive public aid and subsidies. Dufour and Son (2015) identified four potential barriers to open innovation following the literature review: corporate culture, networking, organizational structure and knowledge management systems. They applied the theoretical framework to a single case study to provide insights into how to overcome these barriers by analyzing the various organizational changes undertaken by the company. Bigliardi and Galati (2016) studied the factors hindering the adoption of open innovation in small- and medium-sized companies. Based on the results of a survey they carried out among 157 Italian companies, they identified four main categories of obstacles. Following a cluster analysis, Bigliardi and Galati (2016) found that the main obstacles perceived by knowledge-intensive, high-tech and/or highly innovative industries are those related to the knowledge domain. The cluster of moderately innovative small- and medium-sized enterprises is mainly concerned with financial and strategic issues as well as knowledge barriers. Finally, the cluster composed of companies operating in traditionally less innovative sectors considers significant collaborative and organizational barriers to the adoption of open innovation (Bigliardi and Galati 2016). Ullrich and Vladova (2016a) proposed a weighing and deciding process framework as a possible solution approach to increase the chances of success of open innovation projects. Ullrich and Vladova (2016) found in their research that many studies assess the positive aspects of open innovation processes, while few articles point out the dark sides. However, the best way to encourage open innovation in small- and medium-sized enterprises is to analyze the interdependencies of the two facets and their combined impact on open innovation projects (Bresciani 2017). In recent research by (Yadav et al. 2022), the main challenges in the agri-food supply chain are categorized into five areas: food safety and security, sustainability, technology adoption, food waste and various. These challenges are summarized in Figure 9.5.

Figure 9.5. *Major challenges of agri-food supply chain (source: authors inspired from Yadav et al. (2022))*

9.5. Open innovation in the food supply chain

This section aims to highlight some research works related to the open innovation in food supply chain and the main actors involved in open innovation in this sector.

9.5.1. *Main open innovation research works*

It should be noted that innovation in the agri-food sector generally refers to the addition of a new, unusual ingredient; the design of new product combinations and different processing systems or production procedures, including packaging (Guerrero et al. 2008). Furthermore, this sector has other characteristics such as vertical integration along the industry value chain. Many worldwide players have established a variety of networks leading to technological changes. This practice has promoted cooperation and development of open innovation (Fortuin and Omta 2009; Bresciani 2017). Therefore, Bigliardi and Galati (2016) indicate that agri-food companies try to improve their innovation capacity by adopting an open approach to innovation and by exploiting external knowledge paths to the market. Moreover, recently, the complexity of food supply chains has also been intensified by the need to include multiple sources of external knowledge when researching new successful products and technologies (Miglietta et al. 2017). Several open innovation strategies are undertaken. The objective can range from simple access to external sources of knowledge, to active participation and the creation of inter-organizational knowledge and skills (Miglietta et al. 2017). This is due to the fact that these companies operate in an environment characterized by the dependence of the production process on uncontrollable factors such as the perishability of products and raw materials in an increasingly turbulent market where global competition is active and sustainability requirements are high (Bilali 2018).

In recent years, some researchers and practitioners have been interested in understanding the drivers of innovation, its barriers (Schweisfurth et al. 2011) and the characteristics of new product development in the agri-food sector (Schoen 2017). Others have studied the determinants of innovation (Avermaete et al. 2004), consumer behavior (Moskowitz and Saguy 2013; Schoen 2017), industry-university collaboration (Ivascu et al. 2016) and user participation in innovation processes (Costa and Jongen 2006; Olsen 2015). It should also be noted that the majority of studies on open innovation in the agri-food sector focus on the external process, in which the knowledge base of society is enriched through the external search for knowledge, instead of carrying out profits by transferring ideas to the external environment in a reverse process (Enkel et al. 2009). For example, Devaux et al. (2018) studied the partnership between agribusinesses and other open innovation communities. Other researchers such as Traitler et al. (2011) and Bigliardi and Galati (2016) looked at partnerships with research centers and service providers in the agri-food sector.

Medeiros et al. (2016) explored how agribusiness organizations still need to discover the potential of technical tools and the role of information systems to support open innovation. Gassmann et al. (2010) and Chesbrough (2017) noted the tendency of firms to overlook the possibility of leveraging their created technology and intellectual property in new markets. In an effort to further develop the field of open innovation research in the agri-food sector, Gassmann et al. (2010) found it useful to classify the growing body of literature into nine categories. In 2017, Schoen mapped advances in the agri-food sector with respect to the specific context of the sector by basing its analysis on the proposed open innovation flows. The initial nine perspectives presented by Gassmann and his colleagues have been extended by Schoen to incorporate two new strands important to the agri-food sector. They have been named "system perspective" and "cross-industry/convergence perspective" (Schoen 2017). Furthermore, robust agribusiness firms and their business models have persisted over time (Kanter et al. 2015). They succeeded in overcoming the challenges and adopting appropriate management strategies. Indeed, companies in the agri-food sector are confronted with limits in terms of skills because they are required to position themselves in a new value chain, as in the case of nutraceuticals and functional foods (Sankaran and Mouly 2006; Bigliardi and Filippelli 2022a).

9.5.2. *Actors involved in open innovation*

A food supply chain network is made up of at least three actors: the agri-food company, its suppliers and its consumers (Mentzer et al. 2001). It should be emphasized that customers and consumers play an important role in market innovations. Indeed, food manufacturers largely depend on information obtained from their customers and consumers about their requirements and needs, as well as about the market. Suppliers are also involved in product development (Bigliardi et al. 2020). Collaboration with raw material suppliers often takes the form of cooperative or vertical integration, indicating high collaboration intensity (Bigliardi et al. 2020). Generally, agro-food companies involve actors belonging to the same supply chain when they undertake open innovation projects (Bigliardi and Bottani 2010). Other collaborative partners are identified in the literature (Galati et al. 2016), ranging from universities, research centers, competitors and other industries. In the specific context of food products, previous research has highlighted the key role of some of these actors. More specifically, Wijnhoven et al. (2015) distinguish between a vertical network and a horizontal network. The vertical network is made up of actors involved in the transformation of upstream resources into downstream products (Mentzer et al. 2001; Vorst 2006), namely, the agri-food company, its suppliers and its customers. In addition, distributors can be added to the vertical network (Gellynck and Kühne 2008). Similar considerations apply to open innovation mechanisms: due to the large number of actors involved in the development of innovative food products, innovation activities need to be carefully coordinated (Bigliardi and Galati 2013a, 2013b). Distributors, particularly small specialty stores that typically sell food products, participate in the development of market innovations, such as the use of alternative distribution channels or the extension of the market for food products (Bigliardi et al. 2020). All actors can be involved in organizational innovation, such as forming innovation networks or member research organizations (Bigliardi et al. 2020). The collaborations established between these actors and traditional agri-food companies remain essentially informal and based on long-term relationships (Gellynck and Kühne 2008). In some cases, there is a consortium of artisanal agri-food manufacturers who set the manufacturing rules guaranteeing the quality of the products (Bigliardi et al. 2020). However, networks are not characterized by intensive collaboration. The two main factors hindering collaboration in this context relate to the lack of trust and the lack of means to preserve confidentiality, which both refer to collaborative-type factors preventing the adoption of open innovation, as identified by Bigliardi and Galati (2016). Involving a growing number of actors in the chain to meet the needs of increasingly heterogeneous consumers (Bigliardi and Galati 2016) generates a complex and dynamic system, based on several vertical and horizontal relationships (Sarkar and Costa 2008).

Furthermore, horizontal networks, which are formed by peers or competitors as well as third parties such as government, research institutes and consultants (Schweisfurth et al. 2011), play an important role in the collaboration (Braun 2015). Peers, like competitors, are an important source of new knowledge in the same knowledge domain (Vanhaverbeke 2008). Similarly, third parties and public institutions are important sources of policy and support for innovation (Vanhaverbeke 2008). Furthermore, Gellynck and Kühne (2008) emphasized the essential role of consortia in horizontal networks for the development of food products. Their role is to define production specifications and guarantee quality in order to define the identity of consumers (Bayona et al. 2013).

It should be noted that a limited but growing number of agri-food companies are devoting most of their time and resources to research activities aimed at improving their productivity and their innovations, adopting success factors that go beyond their internal limits (Sarkar and Costa 2008; Bigliardi and Galati 2013). In this context, companies gradually integrate external actors such as research centers, universities and laboratories, intermediate customers and end-users (Mikkelsen et al. 2005) in their innovation processes (Pearson 1990), namely technology push and market pull. With regard to the former, it is mainly about applications of developed technologies to products and inventions without regard to market appeal. Conversely, the market attraction strategy is based on market needs that determine investments in the development of technologies and their integration into products (Khilji et al. 2006). Brem and Voigt (2009) point out that these two strategies are of paramount importance for all types of organizations and that adopting a single strategy can compromise the competitiveness of the organization in the long term. Breuer and Steinhoff (2010) prove that customer orientation is a critical factor for the success of a new product, and also for the success of the company as a whole. However, they point out that the innovation process is often characterized by a lack of customer orientation. Conversely, there is no explicit example of technology push and open innovation in the agri-food sector (Bresciani 2017).

On the contrary, the demand for more personalized and experiential food is growing, with increased attention to customers and personalized, tailored food products (Marsh and Bugusu 2007). In the food market, customer-oriented market niches are changing rapidly, with new needs for product diversification and customization reinforcing the need for innovation across the agri-food sector (Sarkar and Costa 2008). Several companies in the agri-food sector have understood this and have started to open up, in open collaboration, with new co-actors and co-creators in their innovation processes (Bigliardi and Galati 2013), also using sources such as open-innovation sites involving external web users who are usually unrelated to them and participate in their R&D process (Seltzer and Mahmoud 2022). The final frontier of open innovation is crowdsourcing, whereby a company, instead of assigning creative tasks to a disengaged agent, can outsource the creative process to a multitude of website users in the form of a call open to the Internet (Buecheler et al. 2010).

9.6. Open innovation contributions to the food supply chain

The contribution of open innovation for food supply chain is demonstrated through different axes. Companies in the agri-food sector are required to undertake innovations and increase the quality of their products in order to improve consumer confidence and encourage technological development and environmental compatibility (Samadi 2014). Therefore, innovation has become an unavoidable bodily reactivity and the adoption of an effective innovation system to successfully introduce and develop new products in the market has become one of the most important strategies for agribusinesses (Costa and Jongen 2006). To this end, the most recent literature provides studies on open innovation in low-tech industries such as the agri-food sector (Bigliardi et al. 2010; Saguy and Sirotinskaya 2014). It should be noted that food producers are mainly SMEs facing increased competitive pressure, prompting them to seek ways to add value or eliminate costs through the adoption of new technologies, new processes and new practices (Galati and Bigliardi 2019). Furthermore, the predominant nature of innovation is related to demand constraints and conservative consumer behavior (Bayona et al. 2017). Various authors have studied the effect of open innovation practices in companies through case studies (Vanhaverbeke et al. 2008; Bigliardi et al. 2012) or surveys (Lazzarotti et al. 2014; Popa et al. 2017c). These studies show that open innovation is an appropriate new paradigm to address the new challenges facing the agri-food sector (Bayona et al. 2017). It promotes the creation of value throughout the value chain and the establishment of trust and respect. Furthermore, research suggests that inbound open innovation practices can increase a firm's innovation performance (Tomlinson 2010; Huizingh 2011). However, according to Gassmann and Enkel (2004), to get the most out of incoming activities, companies need to understand and integrate external knowledge into their current activities. This is particularly relevant and challenging in the agribusiness sector, given that absorptive capacity is achieved primarily through in-house R&D activities (Cohen and Levinthal 1990; Bayona et al. 2017a). It should be noted that a company interested in innovations is more likely to be involved in the active use of information, but the role of external channels may differ depending on the type of innovation in question (Lee et al. 2010). External sources of innovation such as customers, suppliers, competitors and universities can be considered as the main elements of the research strategy of an agri-food company. However, since the agri-food sector is of low technological potential, it risks hampering the search for external sources of knowledge and further increasing the costs associated with excess information (Laursen and Salter 2006). Open innovation is a paradigm shift in the direction of accelerating the co-development of sustainable innovation, one of the main pillars being the alignment of the entire supply chain on innovations centered on the consumer. This implies that gaps and needs are shared openly and clearly with the innovation partner (Miglietta et al. 2017). In particular, Su et al. (2022) propose some keys factors that can help overcome the challenges of open innovation, which they have classified at three levels (see Figure 9.6).

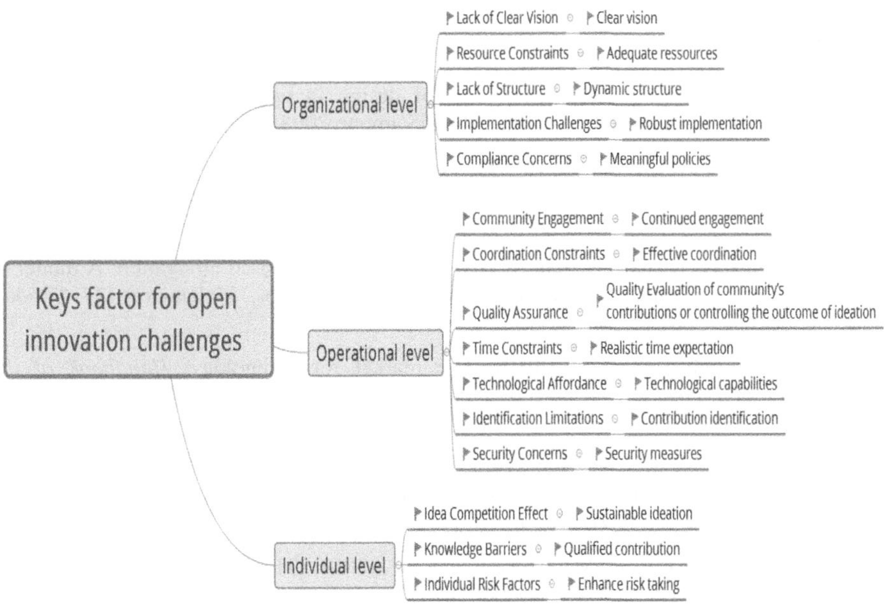

Figure 9.6. *Keys factor to open innovation challenges in food supply chain (source: authors inspired from Yadav et al. (2022)). For a color version of this figure, see www.iste.co.uk/koubaa/optimization.zip*

9.7. Conclusion

This chapter presents an overview of open innovation, the characteristics and challenges of food supply chain and the contribution of open innovation. Companies are challenged to seek the right level of openness to exploit external knowledge related to the practice of open innovation without hampering their innovation effectiveness. However, they are also urged to remain cautious when investing in an open-innovation strategy, especially when efforts invested in external collaborations do not produce meaningful innovation results. Finally, firms are required to seek effective sources of knowledge and exploit them for this purpose. By creating synergies between internal and external knowledge, innovation collaborators could benefit not only from greater innovation efficiency (Greco et al. 2019) but also from their joint sustainable growth (Watson et al. 2018). The main implication for policy makers is the importance that should be given to promoting open innovation by improving internal and collaborative practices. The successful adoption of innovation in the market requires a complex architecture of the innovation process designed to support and push new ideas throughout its life cycle (Silviana 2018).

9.8. References

Acha, V. (2008). Open by design: The role of design in open innovation. *Academy of Management Proceedings*, 1, 1–6. doi: 10.5465/ambpp.2008.33653210.

Ahn, J.M., Minshall, T., Mortara, L. (2017). Understanding the human side of openness: The fit between open innovation modes and CEO characteristics. *R&D Management*, 47(5), 727–740. doi: 10.1111/radm.12264.

Almirall, E., and Casadesus-Masanell, R. (2010). Open versus closed innovation: A model of discovery and divergence. *Academy of Management Review*, 35(January), 27–47. doi: 10.5465/amr.35.1.zok27.

Arora, A. and Gambardella, A. (2010). The market for technology. *Handbook of the Economics of Innovation*, 1, 641–678 [Online]. Available at: https://ideas.repec.org/h/eee/haechp/v1_641.html.

Avermaete, T., Viaene, J., Morgan, E.J., Pitts, E., Crawford, N., Mahon, D. (2004). Determinants of product and process innovation in small food manufacturing firms, *Trends in Food Science & Technology*, 15(10), 474–483.

Bayona, C., Marco, T., Garcia, M., Cruz-Cázares, C. (2013). The impact of open innovation on innovation performance: The case of spanish agri-food firms. *Open Innovation in the Food and Beverage Industry*, December, 74–94. doi: 10.1533/9780857097248.1.74.

Bayona, C., Cruz-Cázares, C., Marco, T., Sánchez, M. (2017). Open innovation in the food and beverage industry. *Management Decision*, 55(April), 526–546. doi: 10.1108/MD-04-2016-0213.

Bertello, A., Bogers, M.L.A.M., De Bernardi, P. (2022). Open innovation in the face of the COVID-19 grand challenge: Insights from the pan-European hackathon "EUvsVirus". *R&D Management*, 52(2), 178–192. doi: 10.1111/radm.12456.

Bigliardi, B. (2019). *Open Innovation and Traditional Food*. Elsevier, Amsterdam.

Bigliardi, B. and Bottani, E. (2010). Performance measurement in the food supply chain: A balanced scorecard approach. *Facilities*, 28(5/6), 249–260. doi: 10.1108/02632771011031493.

Bigliardi, B. and Filippelli, S. (2022a). Open innovation and incorporation between academia and the food industry. In *Innovation Strategies in the Food Industry*, 2nd edition, Galanakis, C.M. (ed.). Academic Press, Cambridge, MA. doi: 10.1016/B978-0-323-85203-6.00003-7.

Bigliardi, B. and Filippelli, S. (2022b). A review of the literature on innovation in the agrofood industry: Sustainability, smartness and health. *European Journal of Innovation Management*, 25(6), 589–611. doi: 10.1108/EJIM-05-2021-0258.

Bigliardi, B. and Galati, F. (2013a). Models of adoption of open innovation within the food industry. *Trends in Food Science & Technology*, 30(1), 16–26. doi: 10.1016/J.TIFS.2012.11.001.

Bigliardi, B. and Galati, F. (2013b). Innovation trends in the food industry: The case of functional foods. *Trends in Food Science & Technology*, 31(2), 118–129. doi: 10.1016/j.tifs.2013.03.006.

Bigliardi, B. and Galati, F. (2016). Which factors hinder the adoption of open innovation in SMEs? *Technology Analysis & Strategic Management*, 28(April), 1–17. doi: 10.1080/09537325.2016.1180353.

Bigliardi, B. and Galati, F. (2018). An open innovation model for SMEs. *Researching Open Innovation in SMEs*, 71–113. doi: 10.1142/9789813230972_0003.

Bigliardi, B., Bottani, E., Galati, F. (2010). Open innovation and supply chain management in food machinery supply chain: A case study. *International Journal of Engineering, Science and Technology*, 2(6). doi: 10.4314/ijest.v2i6.63715.

Bigliardi, B., Dormio, A., Galati, F. (2012). The adoption of open innovation within the telecommunication industry. *European Journal of Innovation Management*, 15(January), 27–54. doi: 10.1108/14601061211192825.

Bigliardi, B., Ferraro, G., Filippelli, S., Galati, F. (2020). The past, present and future of open innovation. *European Journal of Innovation Management*, 24(4), 1130–1161. doi: 10.1108/EJIM-10-2019-0296.

Bogers, M., Zobel, A.-K., Afuah, A., Almirall, E., Brunswicker, S., Dahlander, L., Frederiksen, L., Gawer, A., Gruber, M., Haefliger, S. et al. (2017). The open innovation research landscape: Established perspectives and emerging themes across different levels of analysis. *Industry and Innovation*, 24(1), 8–40. doi: 10.1080/13662716.2016.1240068.

Bogers, M., Chesbrough, H., Moedas, C. (2018). Open innovation: Research, practices, and policies. *California Management Review*, 60(2), 5–16. doi: 10.1177/0008125617745086.

Braun, A. (2015). Linking business model and open innovation – Success and failure of collaborations. *International Journal of Entrepreneurship and Innovation Management*, 19(1–2), 59–76. doi: 10.1504/IJEIM.2015.068440.

Brem, A. and Voigt, K.-I. (2009). Integration of market pull and technology push in the corporate front end and innovation management – Insights from the German software industry. *Technovation*, 29(May), 351–367. doi: 10.1016/j.technovation.2008.06.003.

Bresciani, S. (2017). Open, networked and dynamic innovation in the food and beverage industry. *British Food Journal*, 119(11), 2290–2293. doi: 10.1108/BFJ-08-2017-0458.

Breuer, H. and Steinhoff, F. (2010). Grounded innovation – A research approach for the fuzzy front end of innovation management. *Proceedings of BAI International Conference on Business and Information*, Kitakyushu, 7 [Online]. Available at: https://www.semanticscholar.org/paper/GROUNDED-INNOVATION-%E2%80%93-A-RESEARCH-APPROACH-FOR-THE-Breuer-Steinhoff/3c40f3ac53b5c497adb05e260950f4784ef0398e.

Briones, L.R., Kuch, B., Liu, B.F., Jin, Y. (2011). Keeping up with the digital age: How the American Red Cross uses social media to build relationships. *Public Relations Review*, 37(1), 37–43. doi: 10.1016/j.pubrev.2010.12.006.

Buecheler, T., Sieg, J., Füchslin, R., Pfeifer, R. (2010). Crowdsourcing, open innovation and collective intelligence in the scientific method: A research agenda and operational framework. In *Artificial Life XII – Twelfth International Conference on the Synthesis and Simulation of Living Systems*, 19 August–23 August, Odense, 679–686.

Bullinger-Hoffmann, A., Neyer, A.-K., Koelling, M., Moeslein, K. (2009). Is open innovation really open? A cross-cultural perspective. In *Proceedings of the XIXth ISPIM Conference*, January. International Society for Professional Innovation Management, Vienna.

Carland, J.W., Ensley, M.D., Carland, J.A.C. (1998). The effect of entrepreneurial team skill heterogeneity and functional diversity on new venture performance. *Journal of Business and Entrepreneurship*, 10(1), 1–14.

Carroll, G.P., Srivastava, S., Volini, A.S., Piñeiro-Núñez, M.M., Vetman, T. (2017). Measuring the effectiveness and impact of an open innovation platform. *Drug Discovery Today*, 22(5), 776–785. doi: 10.1016/j.drudis.2017.01.009.

Cassiman, B. and Veugelers, R. (2006). In search of complementarity in innovation strategy: Internal R&D and external knowledge acquisition. *Management Science*, 52(1), 68–82.

Cheng, C.C.J. and Huizingh, E.K.R.E. (2014). When is open innovation beneficial? The role of strategic orientation. *Journal of Product Innovation Management*, 31(6), 1235–1253. doi: 10.1111/jpim.12148.

Chesbrough, H. (2003). *Open Innovation: The New Imperative for Creating and Profiting from Technology*. Harvard Business School Press, Boston.

Chesbrough, H. (2006). Beyond high tech: Early adopters of open innovation in other industries. *R&D Management*, 36, 3. doi: 10.1111/j.1467-9310.2006.00428.x.

Chesbrough, H. (2007). Business model innovation: It's not just about technology anymore. *Strategy & Leadership*, 35(6), 12–17. doi: 10.1108/10878570710833714.

Chesbrough, H. (2017). The future of open innovation. *Research-Technology Management*, 60(1), 35–38. doi: 10.1080/08956308.2017.1255054.

Chesbrough, H. and Appleyard, M.M. (2007). Open innovation and strategy. *California Management Review*, 50(1), 57–76. doi: 10.2307/41166416.

Chesbrough, H. and Bogers, M. (2014a). Explicating open innovation: Clarifying an emerging paradigm for understanding industrial innovation. In *New Frontiers in Open Innovation*, Chesbrough, H., Vanhaverbeke, W., West, J. (eds). Oxford University Press, Oxford.

Chesbrough, H. and Bogers, M. (2014b). Explicating open innovation: Clarifying an emerging paradigm for understanding innovation. In *New Frontiers in Open Innovation*, Chesbrough, H., Vanhaverbeke, W., West, J. (eds). Oxford Academic, Oxford, 3–28 [Online]. Available at: https://papers.ssrn.com/abstract=2427233.

Chesbrough, H. and Brunswicker, S. (2014). A fad or a phenomenon?: The adoption of open innovation practices in large firms. *Research-Technology Management*, 57(2), 16–25. doi: 10.5437/08956308X5702196.

Chin, S.J.I. and Karlsson, M. (2020). The impact of environmental sustainability on innovations in the food industry. Master's Thesis, Lund University, Lund [Online]. Available at: http://lup.lub.lu.se/student-papers/record/9014689.

Christensen, I. and Karlsson, C. (2019). Open innovation and the effects of crowdsourcing in a pharma ecosystem. *Journal of Innovation & Knowledge*, 4(4), 240–247. doi: 10.1016/j.jik.2018.03.008.

Cillo, V., Rialti, R., Bertoldi, B., Ciampi, F. (2019). Knowledge management and open innovation in agri-food crowdfunding. *British Food Journal*, 121(2), 242–258. doi: 10.1108/BFJ-07-2018-0472.

Cohen, W.M. and Klepper, S. (1996). Firm size and the nature of innovation within industries: The case of process and product R&D. *The Review of Economics and Statistics*, 78(2), 232–243. doi: 10.2307/2109925.

Cohen, W.M. and Levinthal, D.A. (1990). Absorptive capacity: A new perspective on learning and innovation. *Administrative Science Quarterly*, 35(1), 128–152. doi: 10.2307/2393553.

Costa, A.I.A. and Jongen, W.M.F. (2006). New insights into consumer-led food product development. *Trends in Food Science & Technology*, 17(8), 457–465. doi: 10.1016/j.tifs.2006.02.003.

Costa, L.B.M., Filho, M.G., Fredendall, L., Paredes, F.J.G. (2018). Lean, six sigma and lean six sigma in the food industry: A systematic literature review. *Trends in Food Science & Technology*, 82(December), 122–133. doi: 10.1016/J.TIFS.2018.10.002.

Covin, J.G. and Slevin, D.P. (1991). Conceptual model of entrepreneurship as firmbehavior. *Entrepreneurship Theory and Practice*, 7, 21–51.

Dahlander, L. and Gann, D.M. (2010). How open is innovation? *Research Policy*, 39(6), 699–709. doi: 10.1016/j.respol.2010.01.013.

De Marco, C.E., Martelli, I., Di Minin, A. (2020). European SMEs' engagement in open innovation when the important thing is to win and not just to participate, what should innovation policy do? *Technological Forecasting and Social Change*, 152, 119843. doi: 10.1016/j.techfore.2019.119843.

Deutsch, C. and McPhee, C. (2013). Editorial: Local open innovation. *Technology Innovation Management Review*, 3(3), 3–5.

Devaux, A., Torero, M., Donovan, J., Horton, D. (2018). Agricultural innovation and inclusive value-chain development: A review. *Journal of Agribusiness in Developing and Emerging Economies*, 8(1), 99–123. doi: 10.1108/JADEE-06-2017-0065.

Dooley, L. and O'Sullivan, D. (2018). Open innovation within the low-technology SME sector. In *Researching Open Innovation in SMEs*, Vanhaverbeke, W., Frattini, F., Roijakkers, N., Usman, M. (eds). World Scientific, Singapore. doi: 10.1142/9789813230972_0008.

Dosi, G., Nelson, R., Winter, S. (eds) (2000). The nature and dynamics of organizational capabilities. In *Nature & Dynamics of Organizational Capabilities*. Oxford University Press, Oxford. doi: 10.1093/0199248540.001.0001.

Dufour, J. and Son, P.-E. (2015). Open innovation in SMEs – Towards formalization of openness. *Journal of Innovation Management*, 3(3), 3–5, 90–117. doi: 10.24840/2183-0606_003.003_0008.

Dumas, J.S. (2002). User-based evaluations. In *The Human-Computer Interaction Handbook: Fundamentals, Evolving Technologies and Emerging Applications*, Jacko, J.A. and Sears, A. (eds). L. Erlbaum Associates Inc, Hillsdale, NJ.

El Bilali, H. (2018). Relation between innovation and sustainability in the agro-food system. *Italian Journal of Food Science*, 30(2). doi: 10.14674/IJFS-1096.

Enkel, E., Gassmann, O., Chesbrough, H. (2009). Open R&D and open innovation: Exploring the phenomenon. *R&D Management*, 39(4), 311–316. doi: 10.1111/j.1467-9310.2009.00570.x.

Faure, G., Barret, D., Blundo-Canto, G., Dabat, M.-H., Devaux-Spatarakis, A., Le Guerroué, J.L., Marquié, C., Mathé, S., Temple, L., Toillie, A. et al. (2018a). How different agricultural research models contribute to impacts: Evidence from 13 case studies in developing countries. *Agricultural Systems*, 165(September), 128–136. doi: 10.1016/j.agsy.2018.06.002.

Faure, G., Chiffoleau, Y., Goulet, F., Temple, L., Touzard, J.-M. (2018b). *Innovation et développement dans les systèmes agricoles et alimentaires*. Editions Quae, Paris. doi: 10.35690/978-2-7592-2813-3.

Fortuin, F.T.J.M. and Omta, S.W.F. (2009). Innovation drivers and barriers in food processing. *British Food Journal*, 111(8), 839–851. doi: 10.1108/00070700910980955.

Galati, F. and Bigliardi, B. (2019). Redesigning the model of the initiation and evolution of inter-firm knowledge transfer in R&D relationships. *Journal of Knowledge Management*, 23(10), 2039–2066. doi: 10.1108/JKM-05-2018-0326.

Galati, F., Bigliardi, B., Petroni, A. (2016). Open innovation in food firms: Implementation strategies, drivers and enabling factors. *International Journal of Innovation Management*, 20(3), 1650042. doi: 10.1142/S1363919616500420.

Garvin, D.A. (1983). Spin-offs and the new firm formation process. *California Management Review*, 25(2), 3–20. doi: 10.2307/41165001.

Gassmann, O. (2006). Opening up the innovation process: Towards an agenda. *R&D Management*, 36(3), 223–228. doi: 10.1111/j.1467-9310.2006.00437.x.

Gassmann, O. and Enkel, E. (2004). Towards a theory of open innovation: Three core process archetypes. *R&D Management Conference*, 21–24 June, Lisbon, 1–18.

Gassmann, O., Enkel, E., Chesbrough, H. (2010). The future of open innovation. *R&D Management*, 40(3), 213–221. doi: 10.1111/j.1467-9310.2010.00605.x.

Gawer, A. and Cusumano, M.A. (2014). Industry platforms and ecosystem innovation. *Journal of Product Innovation Management*, 31(3), 417–433. doi: 10.1111/jpim.12105.

Gellynck, X. and Bianka, K. (2008). Innovation and collaboration in traditional food chain networks. *Journal on Chain and Network Science*, 8(January), 121–129. doi: 10.3920/JCNS2008.x094.

Gianni, M., Gotzamani, K., Vouzas, F. (2017a). Food integrated management systems: Dairy industry insights. *International Journal of Quality & Reliability Management*, 34(2), 194–215. doi: 10.1108/IJQRM-05-2015-0076.

Gianni, M., Gotzamani, K., Vouzas, F. (2017b). Food integrated management systems: Dairy industry insights. *International Journal of Quality & Reliability Management*, 34(January), 194–215. doi: 10.1108/IJQRM-05-2015-0076.

Godfray, C., Beddington, J., Crute, I., Haddad, L., Lawrence, D., Muir, J., Pretty, J., Robinson, S., Thomas, S., Toulmin, C. (2010). Food security: The challenge of feeding 9 billion people. *Science (New York, N.Y.)*, 327(February), 812–818. doi: 10.1126/science.1185383.

Grainger, M. (2010). World summit on food security (UN FAO, Rome, 16–18 November 2009). *Development in Practice*, 20(6), 740–742. doi: 10.1080/09614524.2010.491540.

Greco, M., Grimaldi, M., Cricelli, L. (2019). Benefits and costs of open innovation: The BeCO framework. *Technology Analysis & Strategic Management*, 31(1), 53–66. doi: 10.1080/09537325.2018.1484442.

Grunert, K.G., Jensen, B., Sonne, A.-M., Brunsø, K., Byrne, D., Clausen, C., Friis, A., Holm, L., Hyldig, G., Kristensen, N.H. et al. (2008). User-oriented innovation in the food sector: Relevant streams of research and an agenda for future work. *Trends in Food Science & Technology*, 19(11), 590–602. doi: 10.1016/j.tifs.2008.03.008.

Guerrero, L., Guàrdia, M.D., Xicola, J., Verbeke, W., Vanhonacker, F., Zakowska-Biemans, S., Sajdakowska, M., Sulmont-Rossé, C., Issanchou, S., Contel, M. et al. (2008). Consumer-driven definition of traditional food products and innovation in traditional foods. A qualitative cross-cultural study. *Appetite*, 52(2), 345–354. doi: 10.1016/j.appet.2008.11.008.

Gupta, A.K., Dey, A.R., Shinde, C., Mahanta, H., Patel, C., Patel, R., Sahay, N., Sahu, B., Vivekanandan, P., Verma, S. et al. (2016). Theory of open inclusive innovation for reciprocal, responsive and respectful outcomes: Coping creatively with climatic and institutional risks. *Journal of Open Innovation: Technology, Market, and Complexity*, 2(1), 16. doi: 10.1186/s40852-016-0038-8.

von Hippel, E. (2005). Democratizing innovation: The evolving phenomenon of user innovation. *Journal Für Betriebswirtschaft*, 55(1), 63–78. doi: 10.1007/s11301-004-0002-8.

Howells, G. (2005). The potential and limits of consumer empowerment by information. *Journal of Law and Society*, 32(3), 349–370. doi: 10.1111/j.1467-6478.2005.00328.x.

Huizingh, E.K.R.E. (2011). Open innovation: State of the art and future perspectives. *Technovation, Open Innovation – ISPIM Selected Papers*, 31(1), 2–9. doi: 10.1016/j.technovation.2010.10.002.

Isomäki, A. (2018). Open innovation – What it is and how to do it [Online]. Available at: https://www.viima.com/blog/open-innovation.

Ivascu, L., Cirjaliu, B., Draghici, A. (2016). Business model for the university-industry collaboration in open innovation. In *Procedia Economics and Finance, 3rd Global Conference on Business, Economics, Management and Tourism*, 39(January), 674–678. Elsevier, Rome. doi: 10.1016/S2212-5671(16)30288-X.

Jamett, I., Alvarado, L., Maturana, S. (2017). Analysis of the state of the art of open innovation: Practical implications in engineering. *Revista Ingeniería de Construcción*, 32(2), 73–84. doi: 10.4067/S0718-50732017000200006.

Johnston, A. (2020). Open innovation and the formation of university – Industry links in the food manufacturing and technology sector. *European Journal of Innovation Management*, 24(1), 89–107 doi: 10.1108/ejim-06-2019-0163.

Kafel, P. (2013). Utilisation of quality management methods and tools in food sector organizations. *Zywnosc: Nauka, Technologia, Jakosc*, 1(March), 204–216. doi: 10.15193/zntj/2013/86/204-216.

Kafel, P. (2016). Benefits of management systems integration. *Studia Oeconomica Posnaniensia*, 122–133.

Kafel, P. and Sikora, T. (2014). The level of management maturity in the Polish food sector and its relation to financial performance. *Total Quality Management & Business Excellence*, 25(5–6), 650–663 [Online]. Available at: https://www.tandfonline.com/doi/abs/10.1080/14783363.2013.876182.

Kanter, R., Walls, H., Tak, M., Roberts, F., Waage, J. (2015). A conceptual framework for understanding the impacts of agriculture and food system policies on nutrition and health. *Food Security*, 7, 767–77. doi: 10.1007/s12571-015-0473-6.

King, T., Cole, M., Farber, J.M., Eisenbrand, G., Zabaras, D., Fox, E.M., Hill, J.P. (2017). Food safety for food security: Relationship between global megatrends and developments in food safety. *Trends in Food Science & Technology*, 68(October), 160–175. doi: 10.1016/j.tifs.2017.08.014.

Knickel, K., Brunori, G., Rand, S., Proost, J. (2009). Towards a better conceptual framework for innovation processes in agriculture and rural development: From linear models to systemic approaches. *The Journal of Agricultural Education and Extension*, 15(2), 131–146. doi: 10.1080/13892240902909064.

Kumar, A., Mangla, S.K., Kumar, P. (2022). An integrated literature review on sustainable food supply chains: Exploring research themes and future directions. *Science of the Total Environment*, 821(May), 153411. doi: 10.1016/j.scitotenv.2022.153411.

Kutvonen, A. (2011). Strategic application of outbound open innovation. *European Journal of Innovation Management*, 14(October), 460–474. doi: 10.1108/14601061111174916.

Labrouche, G. (2016). Stratégie d'innovation et capacités dynamiques relationnelles : le cas de l'industrie pharmaceutique. PhD Thesis, Toulouse University Capitole, Toulouse [Online]. Available at: http://www.theses.fr/2016TOU10018.

Läpple, D., Renwick, A., Thorne, F. (2015). Measuring and understanding the drivers of agricultural innovation: Evidence from Ireland. *Food Policy*, 51(February), 1–8. doi: 10.1016/j.foodpol.2014.11.003.

Laursen, K. and Salter, A. (2006). Open for innovation: The role of openness in explaining innovation performance among U.K. manufacturing firms. *Strategic Management Journal*, 27(2), 131–150. doi: 10.1002/smj.507.

Lawrence, A. (2017). Adapting through practice: Silviculture, innovation and forest governance for the age of extreme uncertainty. *Forest Policy and Economics*, 79, 50–60.

Lawrence, R.C., Heap, H.A., Gilles, J. (1984). A controlled approach to cheese technology. *Journal of Dairy Science*, 67(8), 1632–1645. doi: 10.3168/jds.S0022-0302(84)81486-1.

Lazzarotti, V., Garcia, M., Manzini, R., Garcia, M. (2014). Open innovation strategies in the food and drink industry: Determinants and impact on innovation performance. *International Journal of Technology Management*, 66(January), 212–242. doi: 10.1504/IJTM.2014.064588.

Lee, S., Park, G., Yoon, B., Park, J. (2010). Open innovation in SMEs – An intermediated network model. *Research Policy*, 39(March), 290–300. doi: 10.1016/j.respol.2009.12.009.

Lee, S., Olson, D., Trimi, S. (2012). Co-innovation: Convergenomics, collaboration, and co-creation for organizational values. *Management Decision*, 50(May), 817–831. doi: 10.1108/00251741211227528.

Lestari, S.D., Leon, F.M., Widyastuti, S., Brabo, N.A., Putra, A.H.P.K. (2020). Antecedents and consequences of innovation and business strategy on performance and competitive advantage of SMEs. *The Journal of Asian Finance, Economics and Business*, 7(6), 365–378. doi: 10.13106/jafeb.2020.vol7.no6.365.

Lichtenthaler, U. (2017). Open innovation: Past research, current debates, and future directions. *Academy of Management Perspectives*, 25(1). doi: 10.5465/amp.25.1.75.

Liu, Y. and Vrontis, D. (2017). Emerging-market firms venturing into advanced economies: The role of context. *Thunderbird International Business Review*, 59(3), 255–61. doi: 10.1002/tie.21900.

Long, T.B., Blok, V., Dorrestijn, S., Macnaghten, P. (2020). The design and testing of a tool for developing responsible innovation in start-up enterprises. *Journal of Responsible Innovation*, 7(1), 45–75. doi: 10.1080/23299460.2019.1608785.

Lopes, A.P.V.B.V. and de Carvalho, M.M. (2018). Evolution of the open innovation paradigm: Towards a contingent conceptual model. *Technological Forecasting and Social Change*, 132(C), 284–298.

Lopes, C.M., Scavarda, A., Hofmeister, L.F., Thomé, A.M.T., Vaccaro, G.L.R. (2017). An analysis of the interplay between organizational sustainability, knowledge management, and open innovation. *Journal of Cleaner Production Towards a Sustainable Transition*, 142(January), 476–488. doi: 10.1016/j.jclepro.2016.10.083.

March, J.G. (1991). Exploration and exploitation in organizational learning. *Organization Science*, 2(1), 71–87.

Marsh, K. and Bugusu, B. (2007). Food packaging – Roles, materials, and environmental issues. *Journal of Food Science*, 72(3), R39–R55. doi: 10.1111/j.1750-3841.2007.00301.x.

Martinez, M. and Briz, J. (2000). Innovation in the Spanish food & drink industry. *The International Food and Agribusiness Management Review*, 3(2), 155–176. doi: 10.1016/S1096-7508(00)00033-1.

Martínez, C. and Guadalupe, M. (2010). Strategic value of collaboration agreements for the acquisition of open innovation knowledge: An analysis of the ICT sector in Spain. *Contaduría y Administración*, 232(December), 41–64.

Medeiros, G., Binotto, E., Caleman, S., Florindo, T. (2016). Open innovation in agrifood chain: A systematic review. *Journal of Technology Management & Innovation*, 11(3), 108–116.

Mentzer, J.T., DeWitt, W., Keebler, J.S., Min, S., Nix, N.W., Smith, C.D., Zacharia, Z.G. (2001). Defining supply chain management. *Journal of Business Logistics*, 22(2), 1–25. doi: 10.1002/j.2158-1592.2001.tb00001.x.

Meynard, J.-M. (2017). L'agroécologie, un nouveau rapport aux savoirs et à l'innovation. *OCL*, 24(3), D303. doi: 10.1051/ocl/2017021.

Miglietta, N., Battisti, E., Campanella, F. (2017). Value maximization and open innovation in food and beverage industry: Evidence from US market. *British Food Journal*, 119(11), 2477–2492. doi: 10.1108/BFJ-04-2017-0213.

Mikkelsen, B., Kristensen, N., Nielsen, T. (2005). Innovation processes in large-scale public foodservice – Case findings from the implementation of organic foods in a Danish county. *Journal of Foodservice Business Research*, 8(2), 87–105. doi: 10.1300/J369v08n02_07.

Millier, P. (2004). Structuration du champ du management de la technologie et de l'innovation. Working Paper, Cahiers de recherche – EM Lyon, 2003/08.

Minarelli, F., Raggi, M., Viaggi, D. (2015). Innovation in European food SMEs: Determinants and links between types. *Bio-Based and Applied Economics*, 4(1), 33–53. doi: 10.13128/BAE-14705.

Minshall, T., Mortara, L., Napp, J.J. (2007). Implementing open innovation: Challenges in linking strategic and operational factors for HTSFs working with large firms. In *15th High Tech Small Firms Conference*, 14–15 June. Manchester Business School, Manchester.

Misra, A. and Mention, A.-L. (2021). Exploring the food value chain using open innovation: A bibliometric review of the literature. *British Food Journal*, 124(6). doi: 10.1108/BFJ-04-2021-0353.

Moskowitz, H.R. and Sam Saguy, I. (2013). Reinventing the role of consumer research in today's open innovation ecosystem. *Critical Reviews in Food Science and Nutrition*, 53(7), 682–693. doi: 10.1080/10408398.2010.538093.

Mueller, J. and Abecassis-Moedas, C. (2017). Factors influencing the integration of external evaluations in the open innovation process: A qualitative study in micro-firms in the creative industries. *Journal of Strategy and Management*, 10(2), 248–260. doi: 10.1108/JSMA-08-2014-0073.

OECD (2013). Innovation and knowledge management. In *Education Today 2013: The OECD Perspective*. OECD Editions, Paris. doi: 10.1787/edu_today-2013-12-fr.

Ollila, S. and Yström, A. (2016). An investigation into the roles of open innovation collaboration managers. *R&D Management*, 47(2), 236–252. doi: 10.1111/radm.12197.

Olsen, N.V. (2015). Design thinking and food innovation. *Trends in Food Science & Technology*, 41(2), 182–187. doi: 10.1016/j.tifs.2014.10.001.

Parkhe, A. (1993). Partner nationality and the structure-performance relationship in strategic alliances. *Organization Science*, 4(2), 301–324. doi: 10.1287/orsc.4.2.301.

Pearson, A.W. (1990). Innovation strategy. *Technovation*, 10(3), 185–192. doi: 10.1016/0166-4972(90)90023-D.

Piller, F. and Walcher, D. (2006). Toolkits for idea competitions: A novel method to integrate users in new product development. *R&D Management*, 36(3), 307–318. doi: 10.1111/j.1467-9310.2006.00432.x.

Popa, S., Soto-Acosta, P., Martinez-Conesa, I. (2017b). Antecedents, moderators, and outcomes of innovation climate and open innovation: An empirical study in SMEs'. *Technological Forecasting and Social Change*, 118(May), 134–142. doi: 10.1016/j.techfore.2017.02.014.

Porter, M.E. (1985). Technology and competitive advantage. *Journal of Business Strategy*, 5(3), 60–78. doi: 10.1108/eb039075.

Preschitschek, N., Curran, C., Leker, J. (2011). The importance of access to resources in a setting of industry convergence: The case of agriculture and chemistry. In *2011 Proceedings of PICMET '11: Technology Management in the Energy Smart World (PICMET)*. IEEE, Portland, 1–9.

Pustovrh, A., Rangus, K., Drnovsek, M. (2020). The role of open innovation in developing an entrepreneurial support ecosystem. *Technological Forecasting and Social Change*, 152(March), 119892. doi: 10.1016/j.techfore.2019.119892.

Quiñones, R., Caladcad, J.A., Quiñones, H., Caballes, S.A., Abellana, D.P., Jabilles, E.M., Himang, C., Ocampo, L. (2019). Open innovation with fuzzy cognitive mapping for modeling the barriers of university technology transfer: A Philippine scenario. *Journal of Open Innovation: Technology, Market and Complexity*, 5(4), 94. doi: 10.3390/joitmc5040094.

Randhawa, K., Wilden, R., Hohberger, J. (2016). A bibliometric review of open innovation: Setting a research agenda. *Journal of Product Innovation Management*, 33(6), 750–772. doi: 10.1111/jpim.12312.

Roucan-Kane, M., Gramig, B., Widmar, N., Ortega, D., Gray, A. (2013). U.S. agribusiness companies and product innovation: Insights from a choice experiment conducted with agribusiness executives. *International Food and Agribusiness Management Review*, 16(January), 123–140.

Saguy, I.S. (2022). Food SMEs' open innovation: Opportunities and challenges. In *Innovation Strategies in the Food Industry*, 2nd edition, Galanakis, C.M. (ed.). Academic Press, Cambridge, MA. doi: 10.1016/B978-0-323-85203-6.00004-9.

Saguy, I.S. and Sirotinskaya, V. (2014). Challenges in exploiting open innovation's full potential in the food industry with a focus on small and medium enterprises (SMEs). *Trends in Food Science & Technology*, 38(2), 136–148. doi: 10.1016/j.tifs.2014.05.006.

Saguy, I.S. and Sirotinskaya, V. (2016). Open innovation opportunities focusing on food SMEs. *Innovation Strategies in the Food Industry*, 41–59. doi: 10.1016/B978-0-12-803751-5.00003-9.

Salampasis, D.G., Mention, A.-L., Torkkeli, M. (2015). Trust embeddedness within an open innovation mindset. *International Journal of Business and Globalisation*, 14(1), 32–57. doi: 10.1504/IJBG.2015.066099.

Samadi, S. (2014). Open innovation business model in the food industry: Exploring the link with academia and SMEs. *Journal of Economics, Business and Management*, 2(3), 209–213. doi: 10.7763/JOEBM.2014.V2.126.

Sandel, M.J. (2007). *The Case Against Perfection. Ethics in the Age of Genetic Engineering.* Harvard University Press, Harvard. doi: 10.4159/9780674043060.

Sanders, E. and Stappers, P.J. (2008). Co-creation and the new landscapes of design. *CoDesign*, 4(1), 5–18.

Sankaran, J. and Suchitra Mouly, V. (2006). Value-chain innovation in aquaculture: Insights from a New Zealand case study. *R&D Management*, 36(4), 387–401. doi: 10.1111/j.1467-9310.2006.00441.x.

San-Martín-Albizuri, N. and Rodríguez-Castellanos, A. (2012). A conceptual framework for open innovation processes: Integration, distribution and cooperation in knowledge. *Telos*, 14(1), 83–101.

Sarkar, S. and Costa, A.I.A. (2008). Dynamics of open innovation in the food industry. *Trends in Food Science & Technology*, 19(11), 574–580. doi: 10.1016/j.tifs.2008.09.006.

Schilling, M.A. and Phelps, C.C. (2007). Interfirm collaboration networks: The impact of large-scale network structure on firm innovation. *Management Science*, 53(7), 1113–1126. doi: 10.1287/mnsc.1060.0624

Schoen, A. (2017). Openness and collaboration in the food sector: Mapping the field. *British Food Journal*, 119(11). doi: 10.1108/BFJ-03-2017-0126.

Schumpeter, J.A. (1934). Theory of economic development, an inquiry into profits, capital, credit, interest and the business cycle. *Harvard Economic Studies*, 46, 137–148.

Schweisfurth, T., Raasch, C., Herstatt, C. (2011). Free revealing in open innovation: A comparison of different models and their benefits for companies. *International Journal of Product Development*, 13(March), 95–118. doi: 10.1504/IJPD.2011.038867.

Seltzer, E. and Mahmoudi, D. (2012). Planning in public: Citizen involvement, open innovation, and crowdsourcing. *Journal of Planning Literature*, 28(1), 3–18.

Silviana, B.G. (2018). Open innovation model: Enabling the market uptake of innovation. In *Procedia Manufacturing, 11th International Conference Interdisciplinarity in Engineering, INTER-ENG 2017*, 5–6 October 2017, 22(January), 893–899. Elsevier, Tirgu Mures. doi: 10.1016/j.promfg.2018.03.126.

Specht, K., Zoll, F., Siebert, R. (2016). Application and evaluation of a participatory "Open Innovation" approach (ROIR): The case of introducing zero-acreage farming in Berlin. *Landscape and Urban Planning*, 151, 45–54. doi: 10.1016/J.LANDURBPLAN.2016.03.003.

Spencer, J.W. (2008). The impact of multinational enterprise strategy on indigenous enterprises: Horizontal spillovers and crowding out in developing countries. *Academy of Management Review*, 33(2), April. doi: 10.5465/amr.2008.31193230.

Su, N., Mcguckin, S., Abhari, K. (2022). The limits of open innovation: A literature review and research agenda. In *Proceedings of the 55th Hawaii International Conference on System Sciences*, January 4–7. HICSS, Maui. doi: 10.24251/HICSS.2022.686.

Teece, D.J., Pisano, G., Shuen, A. (1997). Dynamic capabilities and strategic management. *Strategic Management Journal*, 18(7), 509–533. doi: 10.1002/(SICI)1097-0266(199708)18:7<509::AID-SMJ882>3.0.CO;2-Z.

Tomlinson, P.R. (2010). Co-operative ties and innovation: Some new evidence for UK manufacturing. *Research Policy*, 39(6), 762–775. doi: 10.1016/j.respol.2010.02.010.

Traitler, H., Watzke, H.J., Saguy, I.S. (2011). Reinventing R&D in an open innovation ecosystem. *Journal of Food Science*, 76(2), R62–R68. doi: 10.1111/j.1750-3841.2010.01998.x.

Tsimiklis, P. and Makatsoris, H. (2016). An open innovation framework for collaborative food product design & manufacturing. *Journal of Innovation Management*, 3(4), 134. doi: 10.24840/2183-0606_003.004_0009.

Tynnhammar, M. (2017). Open innovation and its definitions. In *The XXVIII ISPIM Innovation Conference – Composing the Innovation Symphony*, 18–21 June. International Society for Professional Innovation Management, Vienna [Online]. Available at: www.ispim.org.

Tynnhammar, M. (2019). *New Waves in Innovation Management Research (ISPIM Insights)*. Vernon Press, Wilmington.

Ullrich, A. and Vladova, G. (2016a). Open innovation and its definitions weighing the pros and cons of engaging in open innovation. *Technology Innovation Management Review*, 6(4) [Online]. Available at: https://www.timreview.ca/article/980.

Van der Vorst, J. (2006). Performance measurement in agri-food supply-chain networks – An overview In *Quantifying the Agri-Food Supply Chain*, Ondersteijn, C.J.M., Wijnands, J.H.M., Huirne, R.B.M., van Kooten, O. (eds). Springer, Dordrecht. doi: 10.1007/1-4020-4693-6_2.

Vanhaverbeke, W. (2008). The interorganizational context of open innovation. In *Open Innovation: Researching a New Paradigm*, Chesbrough H., Vanhaverbeke W., West J. (eds). Oxford University Press, Oxford.

Vanhaverbeke, W., Vrande, V., Cloodt, M. (2008). Connecting absorptive capacity and open innovation. *SSRN Electronic Journal*. doi: 10.2139/ssrn.1091265.

Vrande, V., Jong, J.P.J., Vanhaverbeke, W., Rochemont, M. (2009). Open innovation in SMEs: Trends, motives and management challenges. *Technovation*, 29(6–7), 423–437. doi: 10.1016/j.technovation.2008.10.001.

Watson, R., Wilson, H.N., Smart, P., Macdonald, E.K. (2018). Harnessing difference: A capability-based framework for stakeholder engagement in environmental innovation. *Journal of Product Innovation Management*, 35(2), 254–279. doi: 10.1111/jpim.12394.

West, J. and Bogers, M. (2013). Leveraging external sources of innovation: A review of research on open innovation. *Journal of Product Innovation Management*, 31(4), 814–831. doi: 10.1111/jpim.12125.

West, J. and Gallagher, S. (2006). Challenges of open innovation: The paradox of firm investment in open-source software. *R&D Management*, 36(3), 319–331. doi: 10.1111/j.1467-9310.2006.00436.x.

Wijnands, J.H.M., Bremmers, H.J., van der Meulen, B.M.J., Poppe, K.J. (2008). An economic and legal assessment of the EU food industry's competitiveness. *Agribusiness*, 24(4), 417–439. doi: 10.1002/agr.20167.

Wijnhoven, F., Ehrenhard, M., Kuhn, J. (2015). Open government objectives and participation motivations. *Government Information Quarterly*, 32(1), 30–42. doi: 10.1016/j.giq.2014.10.002.

Wikhamn, B.R. and Wikhamn, W. (2013). Structuring of the open innovation field. *Journal of Technology Management & Innovation*, 8(3), 173–185. doi: 10.4067/S0718-27242013000400016.

Yadav, V.S., Singh, A.R., Gunasekaran, A., Raut, R.D., Narkhede, B.E. (2022). A systematic literature review of the agro-food supply chain: Challenges, network design, and performance measurement perspectives. *Sustainable Production and Consumption*, 29(January), 685–704. doi: 10.1016/j.spc.2021.11.019.

Zach, L., Doyle, M., Bier, V., Czuprynski, C. (2012). Systems and governance in food import safety: A U.S. perspective. *Food Control*, 27(1), 153–162. doi: 10.1016/j.foodcont.2012.03.013.

Zhao, S., Sun, Y., Xu, X. (2015). Research on open innovation performance: A review. *Information Technology and Management*, 17(May). doi: 10.1007/s10799-015-0231-7.

10

The Agri-Food Supply Chain and Logistics Problems Optimization

10.1. Introduction

Food is central to life. If we eliminate everything that contributes to life, the only things left are air, food and water. As mentioned in Maslow's Hierarchy of Needs, which points to the stages of human growth, from the physiological need for food, shelter and safety to the heart's desire for love, belonging and esteem. The first of the lower needs on Maslow's hierarchy are physiological needs. The most basic human survival needs are food and water, sufficient rest, clothing and shelter, overall health and reproduction. Furthermore, Maslow states that these essential physiological needs must be met as people move to the next level of fulfillment (McLeod 2007). The agri-food industry is one of the most essential industries in the world. It affects the economic, environmental and social aspects of countries. The US$ 5 trillion shares of this industry represent 10% of consumer costs, 40% of global employment and have been responsible for 30% of greenhouse gas (GHG) emissions since 2015 (Goedde et al. 2015). In the European Union, this sector includes 117.4 and 137.9 billion euros of imports and exports, respectively. In addition to the significant economic, social and environmental impact and the recent growing productivity of the agricultural sector, there is growing concern about the sector's future ability to provide food on a global scale. Current estimates show that by 2050, calorie and product demand will rise by 70% and 100%, respectively (Goedde et al. 2015). These predicted challenges, along with water scarcity (Alcamo et al. 2007) and desertification (Fang et al. 2008), clearly indicate the need to consider measures to optimize the performance of this sector (Goedde et al. 2015). The scale of the food industry's recent challenges of global food insecurity, on the

Chapter written by Omid BOYER, Nadia SAEIDI, Farshad KAVEH, Negin BERJIS and Shila SHARIFZADEH.

one hand, and food waste, on the other hand, are all factors that emphasize the importance of food supply chain management. For instance, food production processes consume more than 10% of the total US energy budget, about 80% of fresh water and about 50% of US lands. However, more than 50% of all produced food is wasted before or after reaching consumers. This is estimated to cause a loss of more than \$165 billion, 25% of fresh water and huge losses of energy, lands and other resources yearly (Govindan et al. 2014; Zhang et al. 2017). Therefore, the food system is a highly complicated and dynamic network involving multiple agents, a wide range of products and a significant number of processes, spanning from production and manufacturing to logistics and retail operations, due to its transnational nature (Trienekens et al. 2012).

Recently, agri-food supply chain management has evolved towards more common interactions between producers, processors, retailers and customers, rather than separate transactions. Agricultural food trade aggregates supply and demand in space and time throughout the food system, from input supply to agricultural products, collection, processing, packaging, transportation and end-user consumption (Barrett and Mutambatsere 2008; Myers et al. 2010). Many related forces impact the food supply chain (FSC), including globalization, market structure and power, consumer tastes and lifestyles, technological changes and regulations. An example of these effects is food products entering local markets worldwide to meet consumer preferences. Agricultural companies and large retailers use logistics and information technology to gain commercial advantages over local farmers and small retailers. The general stages of the FSC first involve the farmers and producers supplying raw materials, whether they are plants or animals. Secondly, food manufacturers and processors conduct most value-added operations to convert food into commercial products. These processes include, but are not limited to, drying, milling, grinding, cooking and packaging. Thirdly, food wholesalers and distributors manage inventory and logistics for transporting food products between various entities. Fourth, retailers such as public markets, grocers, bakers and supermarkets, which serve as low-end stores, provide access to food for consumers. Finally, consumers include households, food service categories such as restaurants and cafes, hospitality sectors such as hotels, and institutional categories such as educational institutions, hospitals and prisons (Dania et al. 2016; Fadhel 2021).

Today's agri-food supply chain (AFSC) is expected to be flexible in offering a variety of options for the end-consumer, from processed food to fresh produce (Zhu et al. 2018). The AFSC faces information asymmetry issues, lack of industrialization and inadequate management, leading to an inefficient supply chain (Kamble et al. 2020). Moreover, the ever-growing demand for food items due to increased population and wealth has placed extra pressure on natural resources like water, energy and land. In addition, it has resulted in increased market volatility and high food prices (Yadav et al. 2022).

Global competition and the unique qualities of agri-food products require food quality to be maintained across the food chain, from farmers to consumers. The supply chain design can help optimize the flow of goods and products through the supply chain to achieve these goals. The supply chain network (SCN) design problem mainly aims to optimize the network's structure in terms of facility location, product flow between stages, product storage and the distribution to service customers quickly and efficiently (Darbari et al. 2019; Mardan et al. 2019; de Sá et al. 2019). Concerning the design of SCNs, it is vital to consider all three strategic, tactical and operational decisions simultaneously because tactical and operational decisions are deeply influenced by strategic decisions (Ozsen et al. 2008). For instance, the effects of locating facilities on vehicle routing and inventory control are significant (Chao et al. 2019; Zandkarimkhani et al. 2020). Faced with these challenges in making these three types of decisions, mathematical models of supply chain and upgraded supply chain management systems can better handle unpredictable situations and consistently meet the quality requirements for food (Eksoz et al. 2014). Therefore, optimization in the agri-food sector helps in making proper strategic, tactical and operational decisions more effectively than ever. To make sure that this chapter can productively cope with current and future challenges in the agricultural supply chain. In the optimization model, two main categories that have impressive effects on strategic, tactical and operational decisions are suggested:

1) The type of material/product flow through the SCN is directly related to the supply chain's design.

2) Different approaches in the AFSC can affect strategic, tactical and operational decisions.

The proposed categories shown in Figure 10.1 will help readers properly understand mathematical models in this field.

The first category is based on the network structure, which has a notable effect on the efficiency of the AFSC. In this category, three subcategories are used to illustrate the flow of goods through the supply chain, namely forward, reverse and closed-loop networks. The second category offers different approaches that can be considered in the AFSC to address various situations in this field. Regarding this second approach, in my opinion, there is some overlap. For example, a halal supply chain can also be sustainable or innovative (I4), so why separate them? I believe that these subcategories need to be revised. It should be noted that the "halal" issue arises only when dealing with meat or when there is a risk of contamination with alcohol. Therefore, why is it considered within the realm of agri-food?

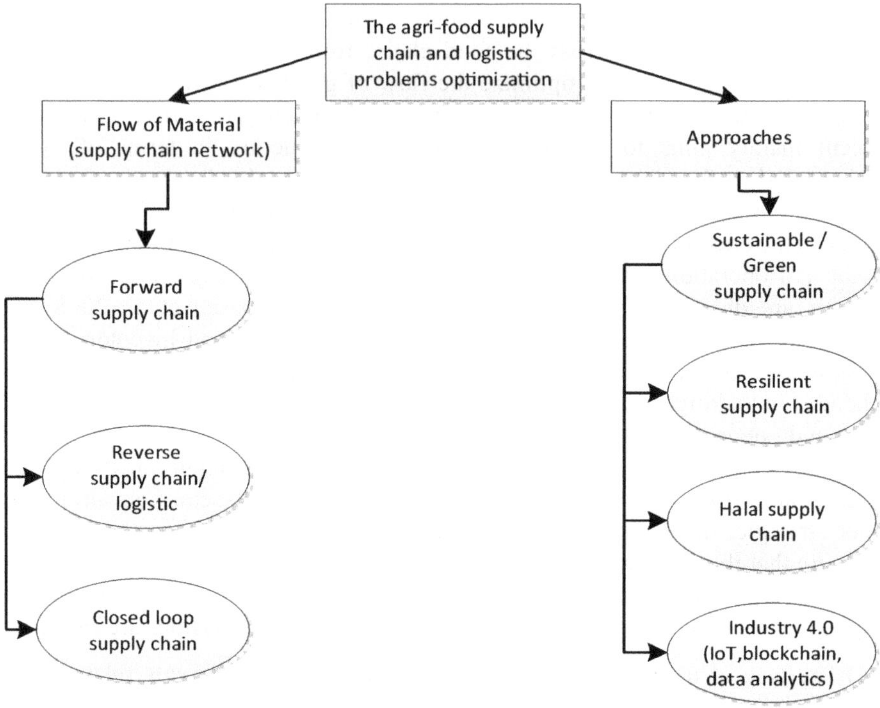

Figure 10.1. *The proposed framework of the chapter*

10.2. Agri-food supply chain network

The SCN includes the necessary centers and connections between these centers to achieve a competitive advantage. In general, SCNs can be divided into three categories: forward, reverse and closed loop. In this section, each type and its role in the FSC are introduced, and finally, a mathematical model for the closed-loop supply chain is examined.

10.2.1. *Forward agri-food supply chain*

A forward supply chain is the network of facilities and activities that perform functions such as product development, procurement of materials from vendors, movement of materials between facilities, manufacturing of products, distribution of finished goods to customers and after-market support for sustainment (Mabert and Venkataramanan 1998). The supply chain includes manufacturers, suppliers, transporters, warehouses, retailers and customers. Within each organization, the supply chain encompasses various functions including, but not limited to, new product development, marketing, operations, distribution, finance and customer service (Chopra 2003).

The general stages of the forward AFSC involve, firstly, farmers and producers who provide raw materials, whether plant- or animal-based. Secondly, food manufacturers and processors where most of the value-added operations to convert food into commercial products take place. Such processes include, but are not limited to, drying, milling, grinding, cooking and packaging. Thirdly, food wholesalers and distributors provide inventory management and logistical operations to transport food products between different entities. Fourthly, retailers such as public markets, grocers, bakers and supermarkets play the role of lower-end outlets that provide food access to consumers. Finally, the consumer stage includes households, food service categories such as restaurants and cafes, the hospitality sector like hotels and institutional categories such as the education sector, hospitals and prisons (Dania et al. 2016; Fadhel 2021). A typical FSC is shown in Figure 10.2.

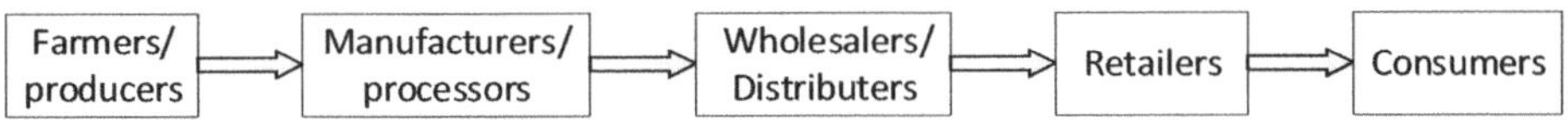

Figure 10.2. *A typical forward agri-food supply chain (Fadhel 2021)*

10.2.2. *Reverse agri-food supply chain*

Usual supply chains often ignore product end-of-life (EOL), leading to products being either misused or wasted during this period. This situation is not in line with current sustainability awareness. Therefore, there is a shift in supply chain design towards incorporating reverse logistics that considers EOL products and aims to handle them in the most eco-friendly manner possible. Reverse logistics (RL) can be defined as a process through which companies can enhance their environmental efficiency by recycling, reusing and reducing materials used (Carter and Ellram 1998). RL also encompasses logistics activities such as product returns, resource reduction, recycling, material replacement, reuse, waste disposal, renovation, repair and reproduction. The standard definition provided by Rogers and Tibben-Lembke (2001) describes RL as the process of planning, implementing and controlling the flow of raw materials, inventory, final goods and related information from the point of consumption back to the origin, facilitating the proper recapturing of value or disposal. RL holds significant importance in today's world due to growing environmental concerns and stringent regulations, as this system can efficiently recycle the substantial amount of waste generated from returned products and transform them into reusable items. RL is an environmentally friendly method that can contribute to greening the supply chain (Govindan et al. 2015). The role of RL in the green supply chain is to reduce environmental damage, address challenges in acquiring production inputs and enhance economic value by facilitating the sale of discarded products in the secondary market. Consequently, it offers a competitive tactical advantage and boosts customer satisfaction (Mangla et al. 2016; Prakash and Barua 2016; Mavi et al. 2017). To establish a reverse flow in supply chains, manufacturers must implement recovery plans that incentivize retailers to collect and return used products (Dekker et al. 2013; Dyckhoff et al. 2013). Jayaraman and Luo (2007) assert that an effective RL system has the potential to enhance profitability by reducing material requirements and bolstering market share through an environmental image. The indirect benefits of an RL program, such as improving corporate image and increasing customer satisfaction, can also add value.

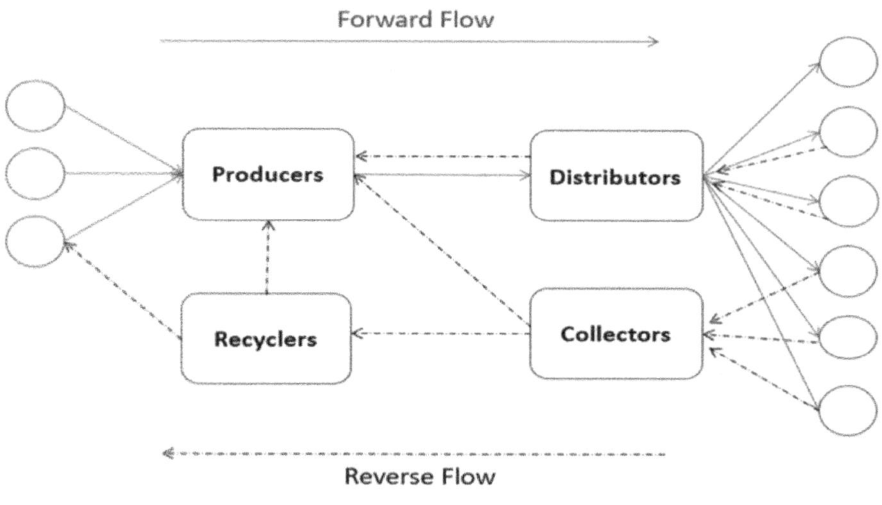

Figure 10.3. *A framework of reverse logistics distribution (Fleischmann et al. 1997). For a color version of this figure, see www.iste.co.uk/koubaa/optimization.zip*

According to Thyberg and Tonjes (2016), food loss and food waste are two influential factors in reverse logistics. Food loss refers to the reduction of food mass across the supply chain, particularly affecting food intended for human consumption, while food waste includes food that was initially produced for human consumption but was either discarded or not used by humans. Waste encompasses foods that have spoiled before being discarded as well as foods that were still edible when discarded. The causes and patterns of food loss and waste are diverse.

Data from the Food and Agriculture Organization of the United Nations (FAO) indicate that food loss is prevalent in developing countries, whereas food waste is a more significant issue in developed countries. The FAO reports that per capita consumer waste in Europe and North America ranges from 95 to 115 kg per year, compared to only 6–11 kg per year in sub-Saharan Africa and South/Southeast Asia.

The higher incidence of food loss in developing countries is often attributed to inadequate infrastructure, limited knowledge and uncertain economic and market conditions (Gogo et al. 2017; Strecker et al. 2022). In contrast, developed countries benefit from advanced infrastructure, extensive expertise and well-defined legal frameworks. However, due to the abundance and affordability of food, families and hospitality services in these countries lack incentives to minimize waste beyond the retail level (Godfray et al. 2010). Figure 10.3 shows how the flow is distributed in a reverse SCN.

10.2.3. *Closed-loop agri-food supply chain*

Resources are becoming scarce, ecosystems are under threat, and the impacts of climate change are significantly affecting the living environment. People are increasingly recognizing environmental pressures, leading to a higher demand for sustainable food production. Food supply chains must embrace innovative technologies that maximize the use of natural resources and materials to enhance their environmental friendliness. The concept of "closing loops", which involves integrating forward and reverse supply chain activities (Guide et al. 2003), is among the options being considered to ensure the sustainability of supply chains (Chaabane et al. 2012). This topic has been extensively researched and has given rise to the field of closed-loop supply chains (CLSCs) (Paksoy et al. 2011). In CLSCs, items that are no longer wanted or used are managed, and their value is (partially) recovered (Flapper et al. 2005). Examples of new logistical structures demonstrate how the concept of closing loops has been applied, leading to improved resource utilization efficiency.

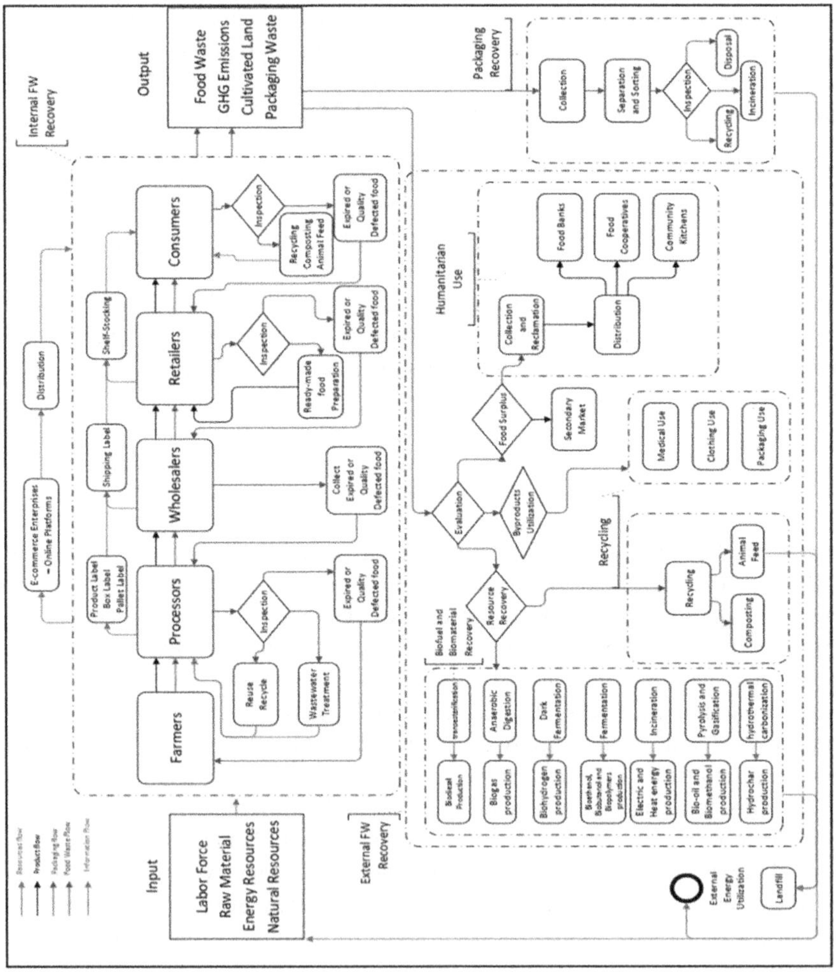

Figure 10.4. The internal and external food waste valorization alternatives (Fadhel 2021). For a color version of this figure, see www.iste.co.uk/koubaa/optimization.zip

A closed-loop agri-food chain is defined as a supply chain that starts from the production and distribution of raw materials, extends to end-users and includes unused products that have lost their value. Food product waste can be repurposed in other supply chains for biofuel or feed production. Unlike the discrete parts industry, in a closed-loop AFSC, the focus is not on regaining the value of the products themselves (within the same chain), but rather on utilizing the components for production as organic matter or a growth medium. Therefore, implementing closed loops in agri-food chains may necessitate a fundamental reassessment of business processes and the redesign of distinct logistical structures (Banasik et al. 2017; Kalantari and Hosseininezhad 2022).

Fadhel (2021) has proposed a comprehensive framework for the agri-food CLSC, as shown in Figure 10.4.

10.2.4. *An example for the closed-loop supply chain*

In this section, the design of a closed-loop logistics network (CLLN) for the citrus fruits supply chain is explained. The logistics network is designed to be multi-period, involving producers, distribution centers, customers, reprocessing centers (compost centers) and compost customers (compost markets).

As depicted in Figure 10.5, goods are transported from producers to customers and distribution centers in the forward flow. In the citrus logistics network design, the forward flow is limited to three periods due to the manufacturers' maximum harvesting period being three months. In addition, in the forward flow, customers receive goods from distribution centers to meet the demand unfulfilled by producers. Distribution centers can dispatch goods in eight periods, aligning with the maximum product and demand period of citrus fruits, which is eight months. Customer locations are assumed to remain fixed.

In the reverse flow, returned products are shipped to vermicomposting centers, where they are converted into organic fertilizer and then distributed to compost customers. Given that producers (gardens) can also be customers of fertilizer, the network can be viewed as a CLSC where producers serve as the same compost customers. Citrus, fresh fruits and vegetable supply chains share similarities and may vary only in the number of periods or distribution type, allowing this network to represent those supply chains with potential minor adjustments (Cheraghalipour et al. 2018).

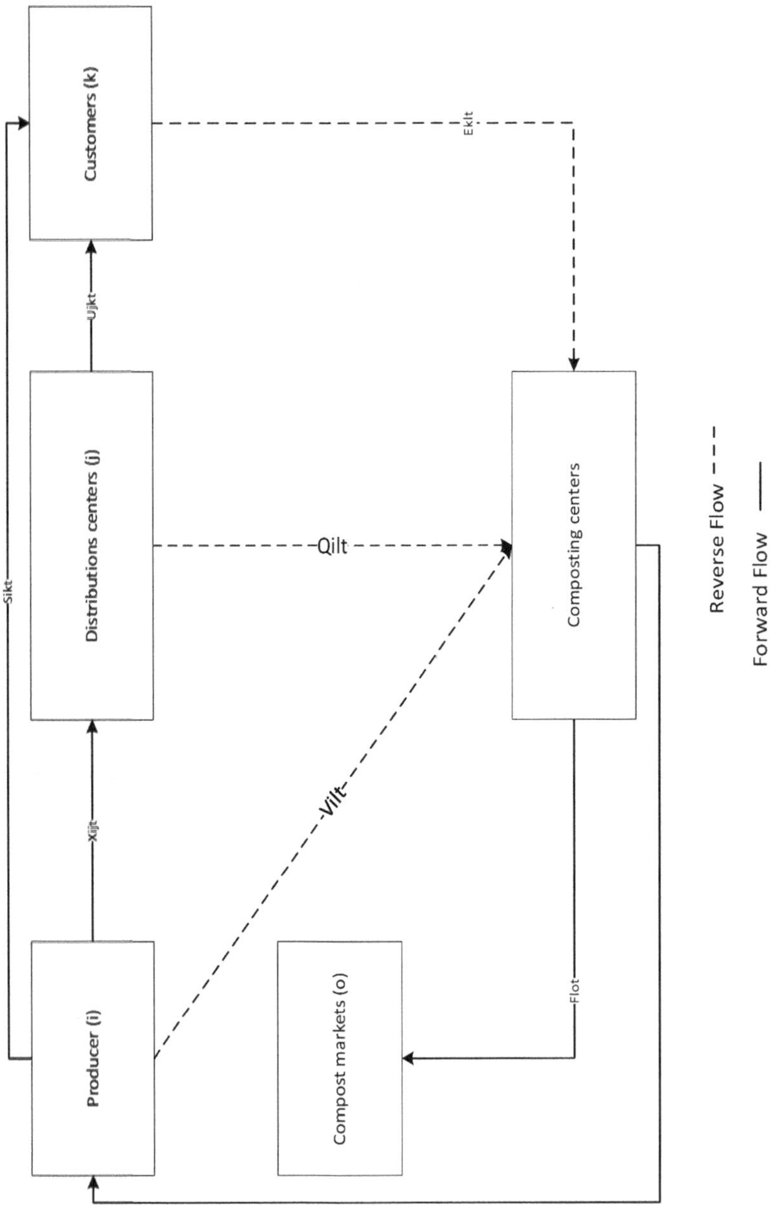

Figure 10.5. *The supply chain network of the problem (Cheraghalipour et al. 2018)*

The model elements including indices, parameters and variables are presented in Table 10. 1.

Indices	
$i = 1, 2, ..., I$	Production locations (gardens)
$j_1 = 1,2, \dots, J_1$	Fixed points of the distribution locations
$j_2 = 1,2, \dots, J_2$	Potential points of the distribution locations
$j = j_1 + j_2$	All points of the distribution locations (sorting)
$k = 1, 2, ..., K$	Customer locations (fruit and vegetable market)
$l_1 = 1,2, \dots, L_1$	Fixed points of the composting locations
$l_2 = 1,2, \dots, L_2$	Potential points of the composting locations
$l = l_1 + l_2$	All points of the composting locations
$t = 1,2,...,t',...T$	Time periods
$o_1 = 1,2, \dots, O_1$	Compost markets
$o_2 = 1,2, \dots, O_2$	Compost customer locations
$o = o_1 + o_2$	
Parameters	
f_i	Fixed cost of opening distribution center j
f_i	The transfer time between retailer i and customer z
Cx_{ij}	Shipping cost per unit of products from producer i to distribution center j
Cs_{ik}	Sipping cost per unit of products from producer i to customer k
Cu_{jk}	Shipping cost per unit of products from distribution center j to customer k
Ce_{kl}	Cost per unit of returned products from customer k to composting center l
Ce_{kl}	Maximum time window for customer
Cq_{jl}	Shipping cost per unit of returned products from distribution center j to composting center l

Cv_{jl}	Shipping cost per unit of returned products from producer i to composting center l
Cf_{lo}	Products from producer composting center l to compost markets o
Ch_t	Holding cost per unit of inventory from distribution centers at time t
Cp_t	Processing and packing cost per unit of products from distribution centers at time t
Cr_t	Compost manufacturing cost per unit of products from composting centers at time t
Cp'	Production cost per unit of products from producers
d_{kt}	Demand of processed product by customer k at time t
λ_{kt}	Maximum production capacity of producer i at time t
λh_j	Holding capacity of distribution center j at time t
λr_j	Compost manufacturing capacity of composting center l at time t
$\propto_t$	Waste percentage of harvested product by producers at time t
θ_t	Waste percentage of stored product by customers at time t
β_t	Waste percentage of stored product by distribution centers at time t
d'_{ot}	Demand of reprocessed product (compost) by compost markets o at time t
ρ	Weighted coefficient (importance) to respond the forward flows
$1-\rho$	Weighted coefficient (importance) to respond the reverse flows
$\emptyset$	Conversion rate of the waste product to reprocessed product
M	A big positive number
Decision variables	
X_{ijt}	Quantity of products shipped from producer i to distribution center j at time t

S_{ikt}	Quantity of products shipped from producer i to customer k at time t
U_{jkt}	Quantity of products shipped from distribution center j to customer k at time t
E_{klt}	Quantity of returned products shipped from customer k to composting center l at time t
Q_{jlt}	Quantity of returned products shipped from distribution center j to composting center l at time t
V_{ilt}	Quantity of returned products shipped from producer i to composting center l at time t
Ih_{jt}	Quantity of stored processed products by distribution center j at time t
λ_{it}	Quantity of production by producer i at time t
W_j	1 if distribution center j is opened at location, 0 otherwise
Y_l	1 if composting center l is opened at location, 0 otherwise
F_{lot}	Quantity of compost shipped from composting center l to compost markets o at time t

Table 10.1. *Notations, parameters and decision variables*

The model has two objective functions: minimization of cost and maximization of responsiveness. The cost objective function considers fixed opening costs, transportation costs, holding costs and production/processing/reprocessing costs. The second objective includes forward responsiveness and reverse responsiveness.

According to the above description, the bi-objective citrus fruits CLLN design can be formulated as follows:

$$MIN\ Z = Z_1 + Z_2 + Z_3 + Z_4 \quad [10.1]$$

$$Z_1 = \sum_{j=1}^{J} f_j + W_j + \sum_{l=1}^{L} f_l * Y_l \quad [10.2]$$

$$Z_2 = \sum_{i=1}^{I} \sum_{j=1}^{J} \sum_{t=1}^{t\prime} Cx_{ij} * X_{ijt} + \sum_{j=1}^{J} \sum_{k=1}^{K} \sum_{t=1}^{T} Cu_{jk} * U_{jkt} + \sum_{i=1}^{I} \sum_{k=1}^{k} \sum_{t=1}^{t'} Cs_{ik} * S_{ikt} + \sum_{i=1}^{I} \sum_{l=1}^{L} \sum_{t=1}^{t'} Cv_{il} * V_{ilt} + \sum_{j=1}^{J} \sum_{k=1}^{L} \sum_{t=1}^{T} Cq_{jl} * Q_{jlt} + \sum_{k=1}^{K} \sum_{l=1}^{L} \sum_{t=1}^{T} Ce_{kl} * E_{klt} + \sum_{l=1}^{L} \sum_{t=1}^{T} \sum_{o=1}^{O} Cf_{lt} * F_{lot} \quad [10.3]$$

$$Z_3 = \sum_{j=1}^{J} \sum_{t=1}^{T} Ih_{jt} * Ch_t \quad [10.4]$$

$$Z_4 = \sum_{i=1}^{I} \sum_{j=1}^{J} \sum_{t=1}^{T} X_{ijt} * Cp_t + \sum_{l=1}^{L} \sum_{t=1}^{T} \sum_{o=1}^{O} F_{lot} * Cr_t + \sum_{i=1}^{I} \sum_{t=1}^{t'} \lambda_{it} * CP' \quad [10.5]$$

$$MaxZ' = p * (\sum_{t=1}^{t'} \sum_{k=1}^{K} \sum_{i=1}^{I} S_{ikt} + \sum_{t=1}^{T} \sum_{k=1}^{K} \sum_{j=1}^{J} U_{jkt})/(\sum_{t=1}^{T} \sum_{k=1}^{K} d_{kt}) + (1-p) * (\sum_{t=1}^{T} \sum_{l=1}^{L} \sum_{o=1}^{O} F_{lot})/ \sum_{t=1}^{T} \sum_{o=1}^{O} d'_{ot} \quad [10.6]$$

$$\lambda_{it} * (1 - \alpha_t) = \sum_{j=1}^{j} X_{ijt} + \sum_{k=1}^{K} S_{ikt} \qquad \forall i \in I, \forall t \in t' \quad [10.7]$$

$$\sum_{i=1}^{I} \sum_{t=1}^{t'} X_{ijt} \leq M * W_j \quad \forall j \in J \quad [10.8]$$

$$\lambda_{it} \leq \lambda c_{it} \qquad \forall i \in I, \forall t \in t' \quad [10.9]$$

$$Ih_{j(t-1)} + \sum_{i=1}^{I} X_{ijt} = Ih_{jt} + \sum_{k=1}^{K} U_{jkt} + \sum_{l=1}^{L} Q_{jlt} \;\; \forall j \in J, \forall t \in T \quad [10.10]$$

$$Ih_{jt} \leq \lambda h_j \quad \forall j \in J, \forall t \in T \quad [10.11]$$

$$\sum_{j=1}^{J} U_{jkt} + \sum_{i=1}^{I} S_{ikt} \leq d_{kt} \; \forall k \in K, \forall t \in T \quad [10.12]$$

$$\sum_{l=1}^{L} V_{ilt} \leq \alpha_t * \lambda_{it} \qquad \forall i \in I, \forall t \in t' \quad [10.13]$$

$$\sum_{t=1}^{t'} \sum_{i=1}^{I} V_{ilt} \leq M * Y_l \quad \forall l \in L \quad [10.14]$$

$$\sum_{l=1}^{L} Q_{jlt} \leq \beta_t * Ih_{j(t-1)} \qquad \forall j \in J, \forall t \in T \quad [10.15]$$

$$\sum_{t=1}^{T} \sum_{j=1}^{J} Q_{jlt} \leq M * Y_l \quad \forall l \in L \quad [10.16]$$

$$(\sum_{l=1}^{I} E_{klt} \leq \theta_t * d_{kt} \quad \forall k \in K, \forall t \in T \quad [10.17]$$

$$\sum_{t=1}^{T}\sum_{k=1}^{K} E_{ilt} \leq M * Y_l \qquad \forall l \in L \qquad [10.18]$$

$$\left(\sum_{i=1}^{I} V_{ilt} + \sum_{j=1}^{J} Q_{ilt} + \sum_{k=1}^{K} E_{ilt}\right) * \emptyset = \sum_{o=1}^{O} f_{lot} \qquad \forall l \in L, t \in T \qquad [10.19]$$

$$\sum_{o=1}^{O} f_{lot} \leq \lambda r_l \qquad \forall l \in L, \forall t \in T \qquad [10.20]$$

$$\sum_{l=1}^{L} f_{lot} \leq d'_{ot} \qquad \forall o \in O, \forall t \in T \qquad [10.21]$$

$$Y_l, W_j \in \{0,1\} \qquad \forall l \in L, \forall j \in J \qquad [10.22]$$

$$X_{ijt}, S_{ikt}, U_{jkt}, V_{ilt}, Q_{jlt}, E_{klt}, F_{lot} \geq 0 \qquad \forall l \in A, v \in N_5, \theta \in \Omega \qquad [10.23]$$

$$Ih_{jt} \geq 0, \lambda_{it} \geq 0 \qquad \forall i \in I, \forall j \in J, \forall t \in T \qquad [10.24]$$

The objective function [10.1] minimizes the total costs, which is comprised of fixed-opening costs [10.2], transportation costs [10.3], holding cost of distribution centers [10.4] and production cost of producers, processing cost of distribution centers and reprocessing costs of composting centers [10.5]. On the contrary, the second objective function [10.6] maximizes the forward and reverse responsiveness of the closed-loop network. As the name implies, responsiveness means satisfying the customers' demands as much as possible. This is done with a balanced equation. Thus, the fraction's numerator shows the shipped products to customers, while the fraction's denominator represents the customers' demands. When the numerator and denominator be equal. Constraint [10.7] ensures that the produced minus the wasted amount equals the number of products shipped from producers to distribution centers and customers. Constraint [10.8] is related to constraint [10.7] and ensures that the products are sent to a potential location only if a distribution center is opened in that location. Constraint [10.9] provides that the product of each producer is less than or equal to the anticipated maximum production rate. This is to consider the production capacity of the producer. Constraint [10.10] ensures that each distribution center's inventory level in each period is equal to the previous period inventory level plus the number of products received from producers minus the number of products shipped to customers and composting centers. Constraint [10.11] shows that distribution center inventory in each period is less than or equal to the holding capacity of the distribution center. Constraint [10.12] states each customer's demand in each period is greater than or equal to the number of products received from producers and distribution centers. Constraint [10.13] shows that returned product shipped to composting centers from each producer is less than or equal to the waste rate of production. Constraint [10.14] expresses that the returned products may be shipped from a distribution center to a composting center only if a composting center is opened in a potential location for such a facility. Similar to constraints [10.13] and [10.14], constraints [10.15] and [10.16] and constraints [10.17] and [10.18] confine the shipped products to the maximum capacity of the facilities and impose the opening of the facility as a pre-condition for shipment of goods. Constraint [10.19] ensures that all returned products from the producers, distribution centers and customers multiplied by the conversion rate equal the total reprocessed product (vermicompost) sent to compost markets. Constraints [10.20] and [10.21] show that the quantity of vermicompost shipped to compost markets in each period is less than or equal to each compost market's manufacturing capacity and demand, respectively. Finally, the binary and non-negativity restrictions on the corresponding decision variables are shown in constraints [10.22]–[10.24].

10.3. Approaches in the agri-food supply chain

Traditional supply chains are changing due to increasing complexity and creating new needs. Different approaches have been proposed to meet current and future requirements. This section introduces different practical approaches in strategic, tactical and operational AFSC decisions.

10.3.1. *Sustainable agri-food supply chain*

Sustainability aims to enhance living standards, not only for the current society but also for future generations. It seeks to strike a balance between the economic, environmental and social impacts of pursuing long-term supply chain objectives. This approach encompasses broader goals beyond merely efficient and profitable production and consumption, including the promotion of socially equitable and environmentally friendly products (Fadhel 2021).

A crucial stage in the journey of food products from farm to fork is their transportation and all associated logistical processes required to deliver food to consumers. Enhancing the efficiency of these processes begins with designing the FSC network, which involves planning processing and distribution centers, managing the cold chain and establishing an RL network. Other tactical considerations related to FSC network design encompass warehouse storage capacity allocation, vehicle routing, material flow optimization, delivery scheduling and inventory management (Fadhel 2021).

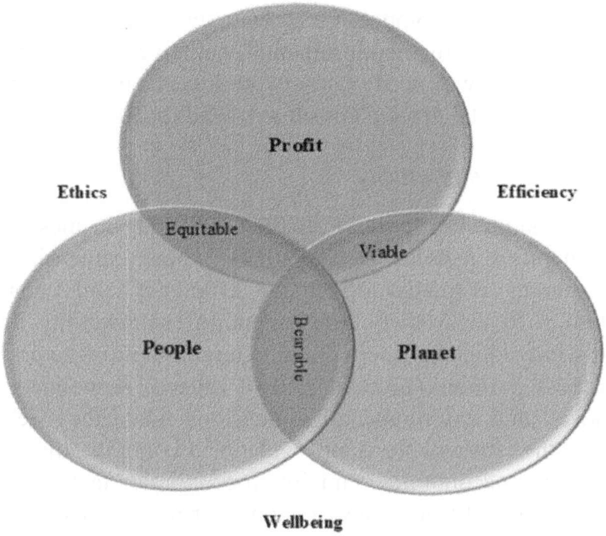

Figure 10.6. *Three bottom lines of sustainability (Fadhel 2021). For a color version of this figure, see www.iste.co.uk/koubaa/optimization.zip*

On the contrary, in response to mounting awareness of the adverse ecological impacts of deforestation resulting from agricultural expansion in tropical regions, civil society organizations, consumers in importing countries and shareholders have started to exert pressure on international suppliers to offer products that adhere to higher social and environmental standards. This has prompted numerous multinational food companies to initiate efforts to "green" their supply chains (Dauvergne and Lister 2012). Globalization holds the potential to alleviate the negative consequences of rising global demand for land-based products by creating incentives for landowners to cultivate without degrading ecosystems or exacerbating socio-economic disparities. However, these initiatives face criticism from those who argue that the standards set do not align with local and social norms for sustainable development (Rueda et al. 2017; Kumar et al. 2022). These recent changes in food supply chain management goals have led to the development of the trending concept: sustainable food supply chain management. As a result, new key performance indicators are developing that can capture the integrated triple bottom line of sustainability, Figure 10.6 in which profit, people and planet are working as drivers of the supply chain decision-making process. The three dimensions of sustainability in the FSC, namely economic, environmental and social dimensions are called the three pillars of sustainability. The sustainability approach of the FSC aims to achieve optimal competitive advantages while maintaining social development and reducing the environmental negative impact. Within each of these three dimensions, we need farmers/producers, manufacturers/processors, wholesalers/distributors, retailers and consumers (Fadhel 2021).

10.3.1.1. *Economical sustainability*

Economic dimensions are critical drivers in food business processes that implement sustainability (Shokri et al. 2014). These dimensions refer to an allocation of resources correctly to achieve efficiency and competitiveness to enhance contribution to the society. Implementing sustainability may lead to an incremental cost while adjusting internal and external facilities to create advantages for all food business partners (Li et al. 2014). Indeed, economic dimensions are influenced by social and environmental dimensions while they become the most critical aspect in the sustainable supply chain. Economic dimensions can be categorized into macroeconomics and microeconomic factors. The economy dimension can be evaluated from macroeconomic or microeconomic points of view in the AFSC. Macroeconomic factors include labor productivity, market concentration and import dependency to support economic growth. Microeconomic factors include cost optimization and revenue management as performance indicators. Moreover, the economic dimensions can be categorized by procurement aspect as inventory management, internal operations aspect as transportation costs and responsiveness efficiency and product development aspect as packaging and product design suitability (Dania et al. 2016). Also, the economic dimension includes consumer preference satisfaction and joint pricing of food products (Fadhel 2021).

10.3.1.2. *Environmental sustainability*

Nowadays, GHGs (including carbon dioxide, nitrous oxide, etc.) and other harmful emissions from the metallurgical industry, fossil fuel combustion and the cement industry not only harm the environment (causing problems such as acid rain, ozone depletion and global warming) but also pose health risks to living organisms. Interestingly, agriculture stands as the most significant source of anthropogenic methane emissions, one of the key GHGs. The FAO warns that these GHG emissions could surge by 30% by 2050 if emission reduction efforts are not intensified. Consequently, the pursuit of green and sustainable development has become imperative for the agriculture industry and the agri-food sector.

Conversely, as the focus on food quality and safety intensifies, the demand for food has shifted towards options that are not only complete but also healthy and environmentally friendly. The consumption of green agri-food has notably risen in both developed and developing nations. Compared to conventional agricultural products, the high nutritional value and reduced risk of diet-related illnesses associated with green agri-food appeal to health-conscious consumers, driving the demand further. Moreover, the growth in green agri-food production can be attributed to its ability to address the sustainability challenges of traditional agriculture, such as soil degradation, nutrient depletion and biodiversity loss, meeting the preferences of environmentally conscious consumers who opt for eco-friendly products.

In the green AFSC, farmers and retailers must invest in diverse green farming technologies and eco-friendly marketing strategies, respectively. For instance, farmers should adopt environmentally sustainable and efficient fertilizers, bio pesticides and smart agricultural machinery powered by renewable energy sources to minimize pollution, albeit potentially increasing their operational costs. Furthermore, retailers have to make additional green marketing efforts to promote the demand as the price of agri-food is usually higher than that of traditional agri-food. Increasing green agri-food consumption will increase revenue, while green farming and marketing efforts will increase costs. Therefore, it is an essential issue for farmers and retailers to determine the reasonable degree of green farming and marketing to balancing the costs and profits, and the concept of green supply chain management came into being to promote the coordinated development of the economy and environment (Cui et al. 2020).

10.3.1.3. *Social sustainability*

Social issues must be considered in all strategic management processes in the supply chain, even though these aspects are the most challenging to measure compared to other aspects since they correlate with intangible elements such as culture, social communities, lifestyle, politics, health, human rights and the aspirations of the communities. The social focus in the food industries can be related to the raw material procurement from the local farmer, vitamins added in the local food and increasing the local income by providing healthy and affordable local products. These social dimensions support community development, labor opportunities and human welfare determined wages, employment and gender ratio to survive in creating fair market share and maintain productivity. Xiao-hui and Jun (2012) added labor standards, life balances, working hours, consumer demands and inflation to become social dimensions indicators. Furthermore, quantitative indicators have been established to measure the social dimensions such as number of employees trained, management levels with specific environment responsibility and number of improvement suggestions submitted by employees (Dania et al. 2016).

From the outset, agricultural and social protection interventions are needed to combat hunger and poverty among poor smallholder farmers. Social protection policies generally aim to reduce socio-economic risks, vulnerability, extreme poverty and deprivation, which helps smallholder farmers improve productivity in crops, fisheries, forestry and livestock and improve access to markers. Social protection can serve as powerful instruments to strengthen access to food, nutrition and essential services, particularly for vulnerable groups in urban and rural settings. Social protection programs can protect food access by increasing purchasing power for those in need. By linking social protection in agricultural settings, AFSC can build resilience and sustainable rural livelihoods. Climate and economic shocks can affect farmers and their production, while increased food price volatility impacts producers and consumers who do not necessarily have the means to cope with them. Therefore, resilience becomes central to the transition towards sustainable agriculture and must be addressed in natural and human dimensions (United Nations Conference on Trade and Development 2020).

10.3.2. *A sustainable food supply chain model example*

The food waste network framework is formulated as a strategic weighted goal programming model that minimizes waste-management costs and maximizes public health and food security while satisfying emissions and energy use constraints. Food waste is generated by multiple stages of the FSC, including farming, processing, retailing and household. Then, the recycled and recovered food waste can be used in humanitarian relief, animal feed, agriculture, or as a source of energy. Input data of different system factors is collected and analyzed to quantify the benefit of valorizing food waste using a specific potential technique. A sustainable benefit model can be used to determine the food waste reduction potentials using input data within the system boundaries. Such models use food waste data to estimate sustainable benefits such as energy reduction and GHG emission reduction.

In addition, the sustainable benefits of RL of food waste are quantified based on geographic distance, valorization characteristics and collection, storage and distribution technologies. Figure 10.7 indicates the network diagram of the food waste. This conceptual design tries to make a clear image of the waste food network (Fadhel 2021).

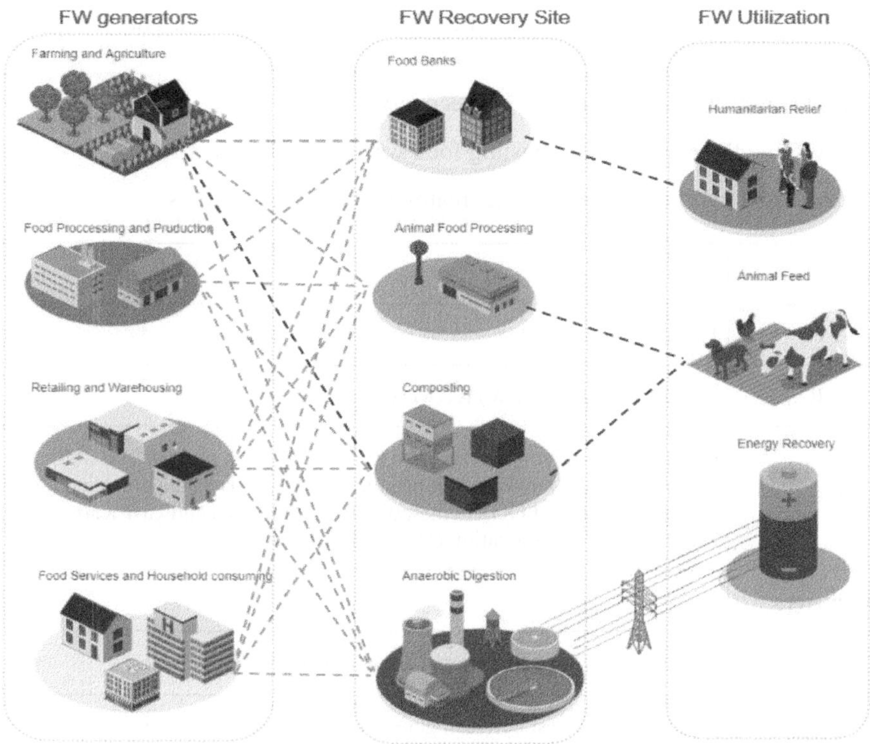

Figure 10.7. *The food waste network diagram (Fadhel 2021). For a color version of this figure, see www.iste.co.uk/koubaa/optimization.zip*

By identifying inputs, outputs and externalities associated with the sustainable food waste reduction system, the following sets and parameters for the food waste network problem are generated, as shown in Table 10.2.

The food waste network framework is formulated as a strategic linear programming (LP) model that aims to minimize total food waste management costs while satisfying emissions and energy use constraints. The formulation is based on the following assumptions:

1) One year of food waste treatment with long-term use of treatment options. This is because establishing treatment facilities requires substantial time and resources, which makes short-term switching infeasible.

2) The food waste is assumed to be separated at the source and ready to be collected by the hauler.

3) The landfill does not involve gas recovery units.

4) Assume the expansion of existing food waste recycling facilities.

5) Assume that the life cycle of the treatment facility is 20 years.

Sets	
G	1,...,G food waste 1
R	1,...,R food waste (FW) recovery sites
$v \in V$	Humanitarian relief (HR), secondary market (SM), animal feed (AF), composting (CO), by-product production (BP), anaerobic digestion (AD), transesterification (TR), incineration (IN), pyrolysis and gasification (PG), hydrothermal carbonization (HC), landfill (LA)
Parameters	
d_{gr}	Distance between FW generator g and recovery site r $(mile)$
tc_{grv}	Transportation cost between generator g and valorization option v at recovery site r ($\$/ton - mile$)
CO_{2rv}	Carbon emissions/absorption resulting from establishing valorization option v in the recovery site r $(ton\ CO_2\ eq/ton)$
pCO_{2v}	Carbon emission resulting from processing FW by option v $(ton\ CO_2$ $eq/ton)$
tCO_{2grv}	Carbon emissions resulting from transport between generator g and option v in site r (ton $CO_2\ eq/ton - mile$)
w_{rv}	Energy required to establish option v in site r (kwh/ton)
pw_v	Energy required to process FW with option v (kwh/ton)
tw_{grv}	Energy required for transport between generator g and option v $(kwh/ton$ $- mile)$
db_r	Development budget allocated for site r (\$)
fc_v	Fixed cost to establish option v ($\$/ton$)

pc_{rv}	Processing cost of FW by option v in site r ($\$/ton$)
cap_{rv}	Capacity of processing facility v in site r (ton)
cp_{gv}	Capacity of generator g (ton) allocated to option v
cp_g	Total capacity of generator g (ton)
ew_v	Power generation resulted from FW treatment by option v (kwh/ton)
e_v	Conversion factor for carbon emission associated with power generation ($ton\ CO_2\ eq/kwh$)
m_v	Conversion factor for carbon emission associated with FW treatment ($ton\ CO_2\ eq/ton$)
fd_v	Sustainability threshold for recovered FW by option v (ton)
max_{LA}	Maximum disposal limit of FW (ton)
Decision variables	
x_{rv}	1 if valorization option v is to be opened in site r, 0 otherwise
y_{grv}	Flow between generator g and recovery site r allocated to valorization option v (ton)

Table 10.2. *Sets and parameters for the food waste network problem*

Given the sets and parameters in Table 10.2, the food waste network model is formulated as follows.

Objective function

$$\min \sum_{r=1}^{R} \sum_{v \in V} fc_v x_{rv} cap_{rv} + \sum_{g=1}^{G} \sum_{r=1}^{R} \sum_{v \in V} d_{gr} tc_{grv} y_{grv} + \sum_{g=1}^{G} \sum_{r=1}^{R} \sum_{v \in V} pc_{rv} y_{grv} \quad [10.25]$$

Demand fulfillment constraint

$$\sum_{g=1}^{G}\sum_{r=1}^{R} y_{grv} \geq fd_v \quad \forall v \in V \qquad [10.26]$$

$$\sum_{r=1}^{R} y_{grv} \leq max_{LA} \quad \forall 1, \dots, G, v: LA \qquad [10.27]$$

Capacity constraints

$$\sum_{g=1}^{G} y_{grv} \leq x_{rv} cap_{rv} \quad \forall r: 1, \dots, R, v \in V \qquad [10.28]$$

$$\sum_{r=1}^{R} y_{grv} \leq cp_{gv} \quad \forall g: 1, \dots, R, v \in V \qquad [10.29]$$

Flow balance constrains

$$\sum_{r=1}^{R}\sum_{v \in V} y_{grv} = cp_g \quad \forall g: 1, \dots, G \qquad [10.30]$$

$$\sum_{r=1}^{R}\sum_{v \in V} x_{rv}\, cap_{rv} \geq \sum_{g=1}^{G} cp_g \qquad [10.31]$$

Development budget constraint

$$\sum_{v \in V} c_v x_{rv} cap_{rv} \leq db_r \quad \forall 1, \dots, R \qquad [10.32]$$

Emissions control constraint

$$\begin{aligned}&\sum_{g=1}^{G}\sum_{r=1}^{R}\sum_{v \in LA} m_v y_{grv} + \sum_{g=1}^{G}\sum_{r=1}^{R}\sum_{v \in V} pCO_{2v} y_{grv} + \\ &\sum_{g=1}^{G}\sum_{r=1}^{R}\sum_{v \in V} d_{gr} tCO_{2grv}\, y_{grv} + \\ &\sum_{r=1}^{R}\sum_{v \in V} CO_{2rv}\, x_{rv} cap_{rv} \leq \\ &\sum_{g=1}^{G}\sum_{r=1}^{R}\sum_{v \in V-\{LA\}} m_v\, y_{grv} + \\ &\sum_{g=1}^{G}\sum_{r=1}^{R}\sum_{v \in \{AD,TR,IN,PG,HC\}} e_v\, ew_v y_{grv}\end{aligned} \qquad [10.33]$$

Energy control constraint

$$\begin{aligned}&\sum_{g=1}^{G}\sum_{r=1}^{R}\sum_{v \in V} pw_v y_{grv} + \sum_{g=1}^{G}\sum_{r=1}^{R}\sum_{v \in V} dg_{rv} y_{grv} + \\ &\sum_{g=1}^{G}\sum_{r=1}^{R}\sum_{v \in LA} d_{gr} tw_{grv}\, y_{grv} + \\ &\sum_{r=1}^{R}\sum_{v \in V} w_{rv}\, x_{rv} cap_{rv} \leq \\ &\sum_{g=1}^{G}\sum_{r=1}^{R}\sum_{v \in \{AD,TR,IN,PG,HC\}} eW_v\, y_{grv}\end{aligned} \qquad [10.34]$$

$$x_{rv} \in \{0,1\}, y_{grv} \geq 0 \quad \forall r: 1, \dots, R, v \epsilon V, g: 1, \dots, G \qquad [10.35]$$

The objective function [10.25] minimizes the fixed, transportation and processing cost of the food waste treatment. The fixed cost represents establishing treatment facilities. The transportation cost is related to transporting food waste from generators to treatment facilities by truck. The processing cost is associated with all activities of food waste treatment by the designated facilities. The set of constraints [10.26] guarantee that the demand for each food waste recovery product is met. Constraint [10.27] limits the food waste disposal amount to a maximum value set by regulators. Constraints [10.28] and [10.29] limit the flow of food waste from generators to treatment sites per the processing capacity of treatment facilities and generators capacities allocated to each treatment facility, respectively. Constraint [10.30] forces the amount of food waste flow within the system to equal the total capacity for each generator. Constraint [10.31] ensures the capacity of all food waste treatment facilities is at least as much as the total capacity of all generators. Constraint [10.32] limits the fixed cost to establish treatment sites per the available budget for each location. Constraint [10.33] controls the net emissions resulting from the food waste treatment system. It ensures that emissions associated with establishing treatment facilities plus emissions from land filling, processing and transportation of food waste must be offset by diverting food waste from disposal to landfill or food waste used for energy recovery. Like constraint [10.33], constraint [10.34] ensures that the energy supply to the food waste treatment system is provided by the energy recovered from the food waste treatment activities. Finally, constraint [10.35] enforces binary values and non-negativity for the decision variables (Fadhel 2021).

10.4. Resilience in the agri-food supply chain

In general, the increasing world population and its urbanization have attracted attention in securing the food supply. Therefore, modern FSCs are among the most vulnerable and fragile since any disruption would rapidly interrupt the operations of the entire network. In addition, any risk that may occur, especially in critical food materials supply, could have terrible consequences on health and safety aspects and arguably affect final customers even more than the FSC itself (Narassima et al. 2024). Food supply chain management has increasingly emphasized resilience in reaction to these discussions on achieving food security on a regional, national or even global level. This is the case in an industry where inventory buffering is already limited due to product perishability, which increases the impact of possible disruptions. Furthermore, there is a reduction in control over the different processing stages due to the globalization of supply chains and a related difficulty in allocating resources for risk mitigation based on the probabilities that the risk will materialize (Bottani et al. 2019).

The incremental frequency and impact of unexpected adverse events have led researchers and practitioners to shift from traditional risk management to resilience (de Sá et al. 2019). Supply chain resilience refers to the supply chain's ability to plan for unforeseen risk occurrences, respond quickly to potential disruptions, return to its original position or move to a better ideal state (Gkountani et al. 2021). Resilience enables systems to cope with the unexpected and ensure continuity of operations and delivery to final customers. Although resilience seems to engage all organizations in a system, it has rarely been investigated how each part contributes to the overall process, such as in the context of supply chains. In other words, resilience is widely defined as the ability of organizations and supply chains to plan for, properly respond to and recover from disruptions in a timely and cost-effective manner; it is the ability to take actions that should return them to an original or perhaps better state than before the disruption. This definition covers the moments before, during and after the disruption and encompasses different phases of resilience: preparedness, response, recovery and growth or adaptation (de Sá et al. 2019). Since organizations are interrelated, their interdependencies augment the consequences of events and cause disruptions across the whole supply chain. To be resilient is a strategy that assists organizations with anticipating disruptions and adapting to new post-event states. However, resilience strategies executed at individual nodes can be detrimental upstream or downstream (Gkountani et al. 2021).

10.4.1. *Risk management and resilience of the agri-food supply chain*

To better understand the great importance and dynamics of supply chain risk management (SCRM), it is critical that three key terms should be defined and analyzed under the supply chain framework, namely risk, vulnerability and resilience. Risk has wide use in many scientific disciplines, and it may refer to failures, disruptions, impacts and/or decisions. One of the most common conceptualizations of risk that includes both qualitative and quantitative components and incorporates the knowledge of a loss-making event, the probability of a loss-making event and the significance (effect) of the event. In comparison to risk, vulnerability is more comprehensive in scope, even wider than integrated SCM, business continuity planning, commercial, corporate risk management or an amalgamation of all of these disciplines. A firm's vulnerability to a disruptive event can be viewed as a combination of the likelihood of disruption and its potential severity. Most approaches to managing risks follow the generic risk management process that consists of three critical stages: risk identification, risk analysis and risk evaluation. In brief, efficient risk management reduces vulnerability by making the supply chain resilient and supply chain management for sustainable food networks relies on the appropriate decision-making about actions that tend to mitigate risk. These actions can be classified into two categories (Keramydas et al. 2015).

10.4.2. *Agri-food supply chain during Covid-19 pandemic*

In December 2019, a strange pneumonia case was detected in Wuhan, China. Initially, any possibility of influenza and other coronaviruses was denied by laboratory testing, but later, the Chinese authority announced it as a novel coronavirus. This virus was named Covid-19 by the WHO. An increasing number of positive cases have been caused by cross-infection, i.e. human-to-human transmission. Due to Covid-19 health precautions, many companies stopped operating to prevent the spread of the virus and to abide by social distancing rules. This greatly affected the operation line throughout the entire supply chain. Limits on the mobility of people have reduced the availability of seasonal workers for planting and harvesting in many countries' food supply chain sectors (OECD 2020). Bottlenecks in transport and logistics have disrupted the movement of products along supply chains. Extra checks (e.g. requirements of new and/or additional certificates) at borders translate to detrimental delays to perishable goods. Some countries quarantined trucks and/or drivers, significantly reducing ground fleets (OECD 2020). Quarantines also applied to ships, which needed to stay longer in port, leading to increased risks of damage and longer delays reaching markets. Trade contraction from Covid-19 has been more profound than the 2008 financial crisis. UNCTAD's latest Global Trade Update (released in October 2020) indicated that the value of international trade in goods declined by about 19% in the second quarter of 2020. Preliminary data for Q3 suggest that global trade growth has remained negative, with a further decline of about 4.5% on a year-on-year basis (see Figure 10.5, UNCTAD 2020). The World Trade Organization (WTO) expects world trade to fall by 13–32% in 2020, 11 while UNCTAD predicts around 20% for the same year (United Nations Conference on Trade and Development 2020).

Due to the prolonged lockdown, most business activities were completely disrupted (Kumar and Kumar 2022) as food is a basic need for everyone. Food production, processing and distribution are also disrupted due to the impact of Covid-19. In general, AFSC is defined as a set of farming, processing/production, testing, packaging, warehousing, transportation, distribution and marketing. All activities of AFSC were disrupted during the period of outbreak of Covid-19, which subsequently led to lockdowns. The FSC covers various items, including the staple to highly perishable commodities. The supply chains of high-value products (perishable products) are most affected by the fast spread of novel coronavirus and the subsequent imposition of lockdowns. The staple product production and harvesting have also been affected to some extent due to the migration of workers. Some of the crops could not be harvested due to workers' shortage, social distancing, transport restriction, etc. (Kumar and Kumar 2021). Covid-19 has also had an instant effect on food systems in developing countries. Limitations to transferring people and goods have impaired access to markets, services and food. Unlike other concurrent crises, rather than threatening the material hardware of food systems, Covid-19 has so far affected the "software" of food systems, highlighting again that connectivity is at the heart of these systems. Drops in demand, the loss of markets and employment, and growing concerns about international cooperation are indications of possible more profound disruptions to come. Amidst this uncertainty, strategies to safeguard the food and nutrition security of the world's poor need to prioritize diversification of products and markets. Nutritious, biofortified crops such as potato, sweet potato, wheat, maize and beans, among others, can play a more significant role in providing necessary critical nutrition at a large scale (Heck et al. 2020).

Changing the pattern of consumers' behavior, transport restriction, import disruption, shortage of laborers, etc. have caused highly-negative shocks for the AFSC (Hobbs 2020). Logistical challenges within supply chains, particularly cross-border and domestic constraints of transportation, as well as labor issues, have resulted in many job losses. High-value commodities, particularly perishables such as fresh fruit and vegetables, meat, fish, milk and flowers, are likely to be affected, resulting to massive food waste as production surpluses declines (Lambert et al. 2020). On the other side, a significant increase in applying online platforms and direct delivery of food items, a major shift in the demand pattern has also been observed. The demand for food in restaurants, hotels decreased rapidly. Labor shortage due to migration is given more importance in the agriculture sector. The labor shortage also influenced the operations of food processing firms and the downstream supply chain. The main reasons for the labor shortage were self-isolation, food security, movement restrictions, migration of workers to their native countries, etc. (Kumar and Kumar 2022).

10.4.3. *An example for the resilient agri-food supply chain*

According to the view of Beamon (1998), the FSC was an integrated process where raw materials were acquired, converted into products and then delivered to the consumer. The chain is linked with the feed-forward flow of materials and the feedback flow of information. Tijskens et al. (2001) concluded that the main fact that differentiates food SCNs from other SCNs is a continuous change in quality from the time the raw materials leave the grower to the time the food reaches the consumer. The hypothetical meat-food SCNs is shown in Figure 10.8, which is defined as consisting of five links that livestock feeding in farms, slaughtered in slaughterhouses, processed meat, then transported to the retailer or supermarket by the distributors to satisfy the consumer demand. Therefore, with increasingly integrated interdependent relationships among all networks, enhancing the anti-disruption ability for the meat-food SCNs is an important issue. Based on the specified meat-food SCNs, the following mathematical model is proposed for system procedures (Jiang et al. 2009).

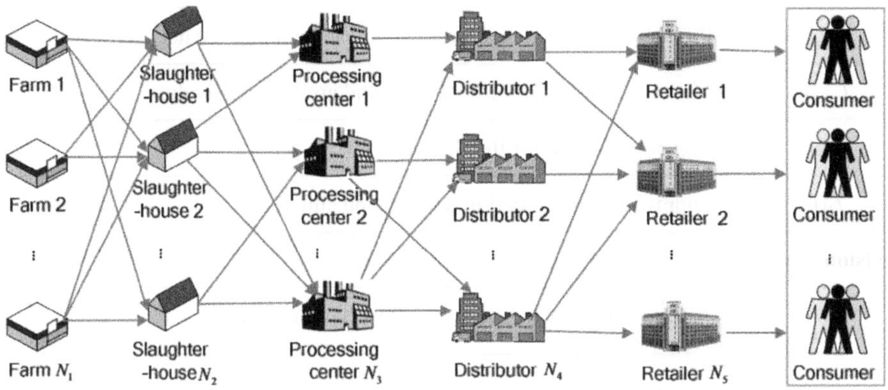

Figure 10.8. *Conceptual framework of meat-food SCNs (Jiang et al. 2009). For a color version of this figure, see www.iste.co.uk/koubaa/optimization.zip*

The mathematical model of resilient meat-food SCNs design problem is presented in this section. The model indices, parameters and decision variables are shown in Table 10.3.

Indices	
Θ	Set of situations
Ω	Set of consumer demand in all demand situations
Parameters	
$D(\theta), P(\theta)$	Amount of consumer demand under situation θ with probability $P(\theta)$
$Y_j^+(\theta), Y_j^-(\theta)$	Shortage or surplus consumer demand under demand situation θ
A	All supply flow arcs of the meat-food SCNs
L	Arc between nodes of the meat-food SCNs, $l \equiv (u, v)$
I_v, O_v	Total input or output degree of node v in meat-food SCNs
a,b	Unit penalty cost for shortage amount and surplus amount, respectively
S_l	Fixed cost for building the arc of supply flow
CP_l, Ct_l	Unit producing cost and unit transportation cost, respectively

λ_v	Process or production coefficient in each node of the networks
L_v , L_l	Capacity constraint for node arc l and arc l, respectively
m_l	Lower constraint for the flow of arc l
$f_l(\theta)$	Supply amount of arc l under demand situation θ
Decision variable	
X_l	1 if arc l are selected as supply flow, 0 otherwise

Table 10.3. *Model indices, parameters and decision variables*

Objective function

$$\min Z = \sum_{l\in A} S_l X_l + \sum_{\theta\in\Omega} P(\theta)\left[\sum_{l\in A}(CP_l + Ct_l) f_l(\theta) + \sum_{l\in A}(aY_j^{+}(\theta), bY_j^{-}(\theta))\right] \quad [10.36]$$

Subject to

$$\sum_{l\in I_v} f_l(\theta) = \lambda_v \sum_{l\in O_v} f_l(\theta) \quad \forall v \in \cup_{i=2}^{4} N_i, \theta \in \Omega \quad [10.37]$$

$$\sum_{l\in I_j} f(\theta)_l + y_j^{+}(\theta) - y_j^{-}(\theta) = \lambda_j D(\theta) \quad \forall j \in N_{5'}, \theta \in \Omega \quad [10.38]$$

$$\sum_{l\in O_v} f_l(\theta) \leq L_v \quad z\forall v \in \cup_{i=1}^{4} N_i, \theta \in \Omega \quad [10.39]$$

$$m_l x_l \leq f_l(\theta) \quad \forall l \in A, \theta \in \Omega \quad [10.40]$$

$$f_l(\theta) \leq L_l * x_l \quad l \in A, \theta \in \Omega \quad [10.41]$$

$$x_l \in \{0,1\} \quad l \in A \quad [10.42]$$

Equation [10.36] defines the total costs, and the first item is the sum of fixed costs for building the arc of supply flow in the meat-food SCN. The second item is the producing costs in nodes and penalty costs of shortage and surplus amount.

Equation [10.37] states the equilibrium of supply flow of the inter-links in the meat-food SCN. Equation [10.38] defines the equilibrium of the final link in the network that the output amount must be equal to the consumer demand under the relaxation conditions (shortage and surplus amount). Equation [10.39] represents the capacity constraint for each supplier in the meat-food SCN. Equation [10.40] denotes the lower amount constraint for the flow of each arc linking two suppliers in the network. Equation [10.41] shows the capacity constraint for the selected arcs that link two suppliers. Equation [10.42] defines the x_l as one if arc 1 is selected as supply flow; otherwise, x_l equals zero.

10.4.4. *Resilient model based on multi-supplier*

As mentioned above, the basic model just minimizes economic costs, whereas the supply chain disruption problem is neglect. Hence, the designed network structure has less resilience when confronted with supply chain failures. Therefore, another network design model considering both the minimization of costs and maximization of resilience under the uncertain demand is constructed as follows:

$$\min \sum_{l\in A} S_l X_l + \sum_{\theta\in\Omega} P(\theta)\left[\sum_{l\in A}(CP_l + Ct_l)e_l(\theta) + \sum_{v\in N_5}(aY_v^+(\theta), bY_v^-(\theta))\right] \quad [10.43]$$

Subject to

$$e_l(\theta) = p_v f_l^n(\theta) \qquad \forall l \in A, \theta \in \Omega \quad [10.44]$$

$$e_l(\theta) \le \rho \sum_{l\in I_v} e_l(\theta) \qquad \forall v \in \cup_{i=2}^{5} N_i, \theta \in \Omega \quad [10.45]$$

$$\sum_{l\in O_v} e_l(\theta) \ge \mu\lambda_v D(\theta) \qquad \forall v \in N_5, \theta \in \Omega \quad [10.46]$$

$$\sum_{l\in I_v} e_l(\theta) = \lambda_v \sum_{l\in O_v} e_l(\theta) \qquad \forall v \in \cup_{i=2}^{4} N_i, \theta \in \Omega \quad [10.47]$$

$$\sum_{l\in I_v} f_l^n(\theta) = \lambda_v \sum_{l\in O_v} f_l^n(\theta) \qquad \forall v \in \cup_{i=2}^{4} N_i, \theta \in \Omega \quad [10.48]$$

$$\sum_{l\in I_{jv}} e_l^n(\theta) + y_v^+(\theta) - y_v^-(\theta) = \lambda_j D(\theta) \qquad \forall v \in N_{5'}, \theta \in \Omega \quad [10.49]$$

$$\sum_{l\in O_v} f_1^n(\theta) \le L_v \qquad \forall v \in \cup_{i=1}^{4} N_i, \theta \in \Omega \quad [10.50]$$

$$e_l(\theta) \le L_l * x_l \qquad l \in A, \theta \in \Omega \quad [10.51]$$

$$m_l x_i \le f_l^n(\theta) \qquad \forall l \in A, \theta \in \Omega \quad [10.52]$$

$$e_l(\theta), y_v^+(\theta), y_v^-(\theta) \geq 0 \quad \forall l \in A, v \in N_5, \theta \in \Omega \qquad [10.53]$$

$$x_l \in \{0,1\} \qquad l \in A \qquad [10.54]$$

Equation [10.44] defines the relationship between expected supply amount and normal-state supply amount in each linkage in the meat-food SCN. Equation [10.45] gives a quantitative formula to select multiple suppliers in each node. Equation [10.46] states the constraint satisfaction level for consumer demand. Equation [10.53] is the non-negative constraint. Other constraints are similar with basic model which can be considered in Liu and Ji (2007).

10.4.5. *Halal supply chain*

The concept of halal has recently received increasing attention. This is due to the fact that the demand for halal food is growing annually as a result of the increase in the number of Muslims worldwide, which is almost 1.8 billion. The halal food market accounts for about 16% of the world food trade, with an estimated trade value of RM 1,628.60 billion (US$ 547 billion) in the global halal market (Yousef 2010). As a result, this trend has attracted the attention of all countries to produce more halal food producers. An important shift in the modern marketing paradigm is that companies are moving away from a consumer-centered approach to a value-based approach, which promotes the concept of marketing to the realm of human ideals, values and spirit (Kotler et al. 2010). Halal (permissible, lawful) is clearly established on values, i.e. Islamic values (Zakaria and Abdul-Talib 2010).

The key to fulfilling the laws that Islam has for identifying haram products is the complete separation of halal products from haram. The most popular forbidden products are alcoholic beverages, pork and fish without scales, which Muslims should not consume. Also, if halal products come in contact with the haram, they can no longer be considered halal and thus become forbidden. Therefore, several countries have established their own halal certification bodies to clearly classify halal. Those institutions get bigger based on the credibility of more people. Some of the representatives of Halal certification offices in the world are Jabatan Kemajuan Islam Malaysia, Singapore's MUIS (Majis Ugama Islam Singapura), Indonesia's MUI (Majelis Ulema Indonesia) and Thailand's CICOT (Central Islamic Council of Thailand) (Othman et al. 2009).

Halal requires a supply chain approach in which the value chain and its supply chain must be perfectly aligned (Assen et al. 2010). This supply chain must live up to the halal promise to the end consumer. On the contrary, the vulnerability of the halal food supply chain (HFSC), the large size and expansion of the halal market and stricter requirements in halal standards have forced brand holders to extend halal towards supply chain management (SCM) (Alam and Sayuti 2011). To make the framework clearer, Figure 10.9 is an example of meat logistics system in Korea in 2019.

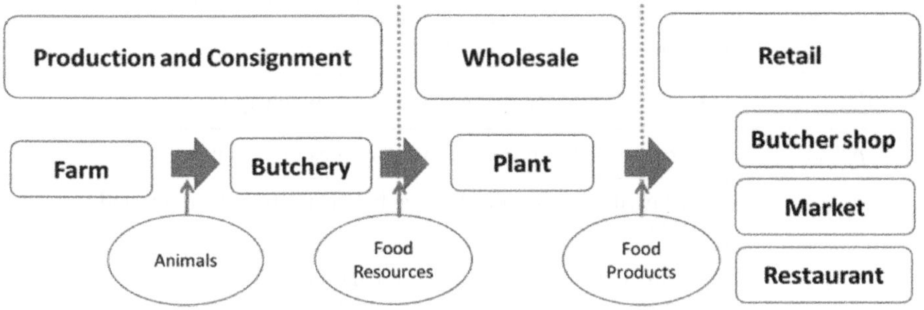

Figure 10.9. *Framework of the meat logistics system of Korea (Kwag and Ko 2019)*

The following is presented a mathematical model of a real halal-logistics network to further clarify supply chain performance with the halal approach (Kwag and Ko 2019). The proposed model includes livestock farms, butcheries that slaughter animals for use as food sources and factories that process food sources for customers' consumption. The entire logistics design is decided based on facilities under contractual terms and operating costs. Notations, parameters and decision variables are shown in Table 10.4.

Notations, parameters	
k	Index for animal types, k ϵ K
K	Set of animal types (K_H = set of halal animals, K_N = set of haram animals) $K_H \cup K_N$
i , i'	Index for farms, i, $i' \epsilon I$
j	Index for butcheries, $j \epsilon J$
m	Index for food resources/product types, $m \epsilon M$
c_{farm}^{ik}	Contract cost for a farm i to provide animal k
$c_{butchery}^{jk}$	Construction cost to build a butchery j that slaughter animal k
c_{plant}^{lm}	Construction cost to build plant l that process food resources m
q_{farm}^{ik}	Maximum capacity of animal k in farm i

$q_{butcheryb}^{jk}$	Maximum capacity that butchery j can slaughter animal k
$q_{butcheryp}^{jm}$	Maximum capacity that butchery j can produce food resources m
q_{plant}^{lm}	Maximum capacity that plant l can produce food resources m
p_{farm}^{ik}	Purchasing cost of animal k
$p_{butchery}^{jk}$	Operation cost to slaughter animal k at butchery j
p_{plant}^{lm}	Operation cost to process food resources m at plant l
D	Least permitted distance to keep between haram facilities and halal facilities
$d_{farm}^{ii'}$	Euclidian distance between farm i and farm i'
$d_{f,b}^{ij}$	Euclidian distance between farm i and butchery j
$d_{b,p}^{jl}$	Euclidian distance between butchery j and plant l
$d_{f,p}^{il}$	Euclidian distance between farm i and plant l
$c_{f,b}^{ijk}$	Transportation cost to move animal k from farm i to butchery j
$c_{b,p}^{jlm}$	Transportation cost to move food resources m from butchery j to plant l
T^{m}	Target demand of food product m
A	A very large positive real number
M_K	Set of food resources m made of animal k
U	Set of farms
U_α	Set of allocated farms
Decision variables	
y_{farm}^{ik}	Allocation decision variable. It will be equal to 1, when animal k from farm i is provided to certain butchery. Otherwise, 0
$y_{butcheryb}^{jk}$	Allocation decision variable. It will be equal to 1, when animal k is slaughtered at butchery j. Otherwise, 0
$y_{butcheryp}^{jm}$	Allocation decision variable. It will be equal to 1, when food resources m is provided at butchery j. Otherwise, 0
y_{plant}^{lm}	Allocation decision variable. It will be equal to 1, when food resources m is produced at plant i. Otherwise, 0
$x_{f,b}^{ijk}$	Network decision variable. It will be equal to 1, when animal k is moved from farm i to butchery j. Otherwise, 0
$x_{b,p}^{jlm}$	Network decision variable. It will be equal to 1, when food resources m is moved from butchery j to plant i. Otherwise, 0

z_{farm}^{i}	Status decision variable. It will be equal to 1, when farm i is halal farm. Otherwise, 0
$z_{butchery}^{j}$	Location decision variable. It will be equal to 1, when butchery j is established. Otherwise, 0
z_{plant}^{l}	Location decision variable. It will be equal to 1, when plant l is established. Otherwise, 0

Table 10.4. *Notations, parameters and decision variables*

Objective function

$$\sum_{i \in I} c_{farm}^{i} \cdot z_{farm}^{i} + \sum_{i \in I} \sum_{k \in K} p_{farm}^{ik} \cdot q_{farm}^{ik} \cdot y_{farm}^{ik} + \sum_{i \in I} \sum_{j \in J} \sum_{k \in K} c_{f,b}^{ijk} \cdot d_{f,b}^{ij} \cdot q_{farm}^{ik} \cdot x_{f,b}^{ijk} + \sum_{j \in J} c_{butchery}^{j} \cdot z_{butchery}^{j} + \sum_{j \in J} \sum_{k \in K} p_{butchery}^{jk} \cdot q_{butcheryb}^{jk} \cdot y_{butcheryb}^{jk} + \sum_{j \in J} \sum_{l \in L} \sum_{m \in M} c_{b,p}^{jlm} \cdot d_{b,p}^{jl} \cdot q_{butcheryp}^{jm} \cdot x_{b,p}^{jlm} + \sum_{l \in L} c_{plant}^{l} \cdot z_{plant}^{l} + \sum_{l \in L} \sum_{m \in M} p_{plant}^{lm} \cdot y_{plant}^{lm} \quad [10.55]$$

Equation [10.55] is the model objective function that minimizes solvent food logistics operations' total network development cost. The equation consists of three different types of cost. The construct costs for farms, butchers and factories are the first types of cost. Then, there are the costs of buying from farms, the operation costs included the cost of slaughtering from butcheries and the cost of producing food products from factories. Finally, shipping costs between farms to butcheries and butchers to factories:

$$z_{farm}^{i} \geq y_{farm}^{ik} \qquad \forall i, \forall k \qquad [10.56]$$

$$z_{butchery}^{j} \geq y_{butcheryb}^{jk} \qquad \forall j, \forall k \qquad [10.57]$$

$$z_{plant}^{l} \geq y_{plant}^{lm} \qquad \forall l, \forall m \qquad [10.58]$$

$$D . z_{farm}^{i'} - d_{farm}^{ii'} \leq A . z_{farm}^{i} \qquad \forall i, \forall i' \qquad [10.59]$$

$$D . \sum_{k \in k_H} y_{butcheryb}^{jk} - d_{f,b}^{ij} \leq A . z_{farm}^{i} \qquad \forall i, \forall j \qquad [10.60]$$

$$D . \sum_{m \in M} y_{plant}^{lm} - d_{f,b}^{il} \leq A . z_{farm}^{i} \qquad \forall i, \forall l \qquad [10.61]$$

$$2.x_{f,b}^{ijk} \leq y_{farm}^{ik} + y_{butcheryb}^{jk} \quad \forall i, \forall j, k \in k_H \qquad [10.62]$$

$$2.x_{b,p}^{jlm} \leq y_{butcheryp}^{im} + y_{plant}^{lm} \quad \forall j, \forall l, \forall m \qquad [10.63]$$

$$\sum_{i \in I} q_{farm}^{ik} . x_{f,b}^{ijk} \geq q_{butchery}^{jk} . y_{butcheryb}^{jk} \quad \forall j, \forall k \qquad [10.64]$$

$$\sum_{i \in I} q_{butchery}^{jm} . x_{b,p}^{jlm} \geq q_{plant}^{lm} . y_{plant}^{lm} \quad \forall l, \forall m \qquad [10.65]$$

$$\sum_{i \in I} q_{plant}^{lm} . y_{plant}^{lm} \geq T^{m} \quad \forall m \qquad [10.66]$$

$$q_{farm}^{ik} \geq y_{farm}^{ik} \quad \forall i, \forall k \qquad [10.67]$$

$$\sum_{k \in k_N} q_{farm}^{ik} . z_{farm}^{i} = o \quad \forall i \qquad [10.68]$$

$$\sum_{j \in J} x_{f,b}^{ijk} \leq 1 \quad \forall i, \forall k \qquad [10.69]$$

$$\sum_{l \in L} x_{b,p}^{jlm} \leq 1 \quad \forall j, \forall m \qquad [10.70]$$

$$y_{butcheryyb}^{ik} = y_{butcheryp}^{jm} \quad \forall i, \forall k, m \in M_k \qquad [10.71]$$

$$y_{farm}^{ik} \in \{0,1\} \quad \forall i, \forall k \qquad [10.72]$$

$$y_{butcheryb}^{jk} \in \{0,1\} \quad \forall j, \forall k \qquad [10.73]$$

$$y_{butcheryp}^{jm} \in \{0,1\} \quad \forall j, \forall m \qquad [10.74]$$

$$y_{plant}^{lm} \in \{0,1\} \quad \forall l, \forall m \qquad [10.75]$$

$$x_{f,b}^{ijk} \in \{0,1\} \quad \forall j, \forall k, \forall i \qquad [10.76]$$

$$x_{b,p}^{jlm} \in \{0,1\} \quad \forall j, \forall k, \forall i \qquad [10.77]$$

$$z_{plant}^{i} \in \{0,1\} \quad \forall j \qquad [10.78]$$

$$z_{butchery}^{j} \in \{0,1\} \quad \forall j \qquad [10.79]$$

$$z_{plant}^{l} \in \{0,1\} \quad \forall l \qquad [10.80]$$

Constraints [10.56]–[10.58] ensure the status of facilities in terms of whether or not they are halal by comparing the relationship between two variables. Constraints [10.59]–[10.61] indicate the distance that facilities must be maintained from forbidden farms to be considered halal. Notation D is the minimum allowable distance between halal and haram facilities. If a specific halal facility gets closer to the shortest allowed distance, it becomes a haram facility. Constraints [10.62] and [10.63] define the relationship between the facilities and supply. These constraints ensure that the facilities receiving the products continue to operate. Constraints [10.64]–[10.66] ensure facilities receive supplies equal to or more than their maximum capacity. This model seeks to develop a design for the halal food logistics network. Therefore, it should be considered when operating with maximum capacity. The constraints indicate that butcheries, factories and consumers must receive enough products at their maximum capacity from the farms, butcheries and factories. Constraint [10.67] ensures that only farms that raise halal animals can be their suppliers. Condition [10.68] shows that if the farm has any haram animal, the farm is known as forbidden. In simpler terms, the constraints can be considered to be the conditions of haram farms that cannot supply animals to butcheries. Constraint [10.69] states that each active farm can supply its animals to only one butchery, while constraint [10.70] shows a similar concept to constraint [10.69] for the supply of food resources from the butchery to the factory. Constraint [10.71] forces the butcheries that receive animals to slaughter them and provide food resources. The cost of waste can happen when many butchers are active, even though they do not slaughter the animals they receive from the farms. The constraint ensures that problems do not occur. Constraints [10.72]–[10.80] set the decision variables to be binary variables.

10.5. New approaches in the agri-food supply chain

Today, SCNs depend on essential technologies that facilitate the integration of planning and implementation of systems, visibility in logistics, smart procurement and warehousing, spare part management and advanced analytics. Therefore, demand management will need to be implemented in more detail using techniques such as mass customization, creative distribution concepts and the complicated practices for scheduling (Soosay and Kannusamy 2018). The Industry 4.0 tools and practices are part of warehouse robotics, road transportation, logistics and supply chain, and the rise of the virtual logistics team (O'Byrne 2020).

The fourth industrial revolution, Industry 4.0, is a new approach that collects different concepts like digitalization, new technologies and interconnectedness. This concept is attracting decision-makers in business and researchers in academia. The first time, the term "Industry 4.0" was introduced by a working group to the German federal government to show a vision of industrial manufacturing's future. After that, it included part of the Action Plan High-Tech Strategy of government in 2020 to ensure technological leadership with digitalization and Internet of Things (Klitou et al. 2017). Industry 4.0 is expected to result in four long-term relationship paradigm shifts and changes to the landscape of European manufacturing: factory and nature, factory and local community, factory and value chains (distributed and responsive manufacturing through collaborative processes, enabling mass customization of products and services) and factory and humans. These will affect technology implementation in the broader manufacturing and distribution environment (Santos et al 2017).

The industry 4.0 revolution in supply chains requires us to adopt new technologies, engaging in capability development and transforming business models and network structures to have good vertical integration. Therefore, it is necessary to change traditional relationships between stakeholders in a supply chain, such as suppliers, manufacturers, wholesalers, retailers and customers. Some industries such as electronics and fast-moving consumer goods are already adopting state-of-the-art technologies and moving along the supply chain 4.0 continuum at present, while other sectors are still lagging. There is concern that the diffusion rate of modern technologies in the agri-food industry has been much slower than anticipated, suggesting that their implementation may involve challenges or lack of collaboration (Soosay and Kannusamy 2018).

Digital platforms mediate the effects of digital transformations on commercial and ecological performance (Li et al. 2020). The digital changes enable the lean production system to decrease waste, support logistical processes and information and use flexible production and safety stock (Hoellthaler et al. 2020). However, digitalization is a crucial component of modern supply chains. Material requirement planning (MRP I&II) implementation can be considered a starting point for digitizing supply chain operations (Kittipanya-Ngam and Tan 2020). Enterprise resource planning systems and continual digital transformations increase efficiency in food production technologies facilitated the downstream food supply chain (Gharehgozli et al. 2017). Also, the new farming-related processes, including machinery, robots and drones, help in last-mile deliveries instead of traditional transportation methods (Robinson 2017). Agri-food 4.0, a digitization approach for sustainable agriculture acceleration and support, land management and competitiveness, the agri-food sector presents an excellent potential for improved efficiency, intelligence, performance and sustainability accompanied by Internet-based networks and services (Hernández Jorge et al. 2018; Dadi et al. 2021).

The following section will explain some new approaches used in the optimization model to convert the traditional food supply chain to a modern one. These new approaches include the big data analytics, blockchain-technology Internet of Things and radiofrequency detection. Also, a mathematical model from the literature is presented to show the applicability of these approaches in the FSC.

10.5.1. *Big data analytics*

Although the advances of computer systems and Internet technologies have witnessed the development of computing hardware following Moore's law for several decades, the problems of large-scale handling data still exist when entering the age of big data. Big data is the type of data that most current information systems or methods cannot process. The most classic data mining method for a centralized data analysis process may not be applied directly to big data. The data size of big data has three features: volume, velocity and variety. Therefore, the data size is large, the data creation is fast and the various type of data collected from different sources. More characteristics of big data are presented in other studies, such as veracity, validity, value, variability, venue, vocabulary and vagueness (Tsai et al. 2015).

The generated data in a supply chain has features of big data. This issue leads to emerging research areas like data science and big data analysis that include statistical and mathematical analysis with supply chain management (Zhao et al. 2017). Using tools and techniques developed by big data analytics to decide supply chains has become widespread. Quick and precious decisions can assist businesses by analyzing and comprehending real-time results that deal with big data. Organizations can also improve the supply chain by cost reduction and risk mitigation (Govindan et al. 2018). Combining data sciences and AFSC is too essential for producing new knowledge by extension of farmers' data, advanced services, processes and software. It also has a critical role in adapting factories-of-future and information and communication technologies (ICT) and is related to agricultural patterns and big data models (Lezoche et al. 2020; Dadi, et al. 2021). Numerous research works use big data analytics in supply chain management in different fields such as procurement, demand forecasting, quality management and inventory control, but few types of research use big data analytics in supply chain optimization models.

10.5.2. *Internet of Things (IoT)*

Information technology (IT) is one of the most critical enablers for effective supply chain management. IT can aid in dealing with the numerous risks in the supply chain because of an ever-changing environment and the growth of sophistication in supply chain management. IT enables the integration of an organization's various internal and external processes with stakeholders. The integration can be achieved by developing communication and collecting and sharing data to increase supply chain performance effectively. The IoT is a modern IT development that can be used in supply chain management. Communication level rise in the supply chain by using the IoT due to independent coordination among things through the supply chain commodities and the possibility of human to things communication. These strengths create new options to deal with more supply chain challenges effectively. Therefore, the IoT leads to a new level of visibility, agility and adaptability in supply chain management to manage new challenges with more effective decision-making (Ellis et al. 2015; Ben-Daya 2019).

Smart farming systems based on IoT and embedded systems that improve food production are taking more attention among people (Mahbub 2020). Agri-food lacks using ICT among different production systems sectors due to the high cost of data accumulating and analytics and using various decision support tools (Brewster et al. 2017; Dahane et al. 2022). Precision agriculture (PA) is an excellent example of the IoT application in agriculture. The IoT can help to have self-adaptive supply chains that can learn and have continued improvement independently (Verdouw et al. 2016; Dadi et al. 2021).

A radiofrequency identification (RFID) system is often seen as a requirement for the IoT. RFID is an automated technology that uses radio waves to aid machines or computers in identifying objects, recording metadata or controlling individual targets. By connecting the RFID reader to the Internet terminal, the readers can identify, track and monitor the objects attached with tags globally, automatically and in real time if needed (Jia et al. 2012). Radiofrequency detection (RFID) is a pervasive technology with complete food logistics applications and cold chain monitoring, which is vital to producing storage retails and supply chain management. It does not need visual contact and is more accurate, providing real-time data consisting of a tag, reader and software (Kumar and Srivastava 2018). It is also used in resolving food safety and security issues, giving an account of food freshness and the origin of food (Athauda and Karmakar 2019).

10.5.3. *Blockchain technology*

Blockchain technology is defined as a "distributed database of records, or public ledger of all transactions or digital events that have been executed and shared among participating parties" (Crosby et al. 2016). This technology can potentially disrupt the supply chain because of its reliability, information legitimacy and capability to enable smart contractual relationships (Saberi et al. 2019). Although there is a growth in blockchain usage in recent years, this technology has some issues such as behavioral, organizational, technological or policy-oriented barriers to execute in the SCN (Crosby et al. 2016; Queiroz and Wamba 2019; Saberi et al. 2019). The old supply chains encounter issues such as lack of trust and complexity in different centers when the value exchange involves massive paper settlements. Transfer costs and transactions are vulnerable to fraud involving intermediaries increases the supply chain operation cost by 70% of the final product cost (Kamilaris et al. 2019). Blockchain technology implementation in the supply chain improves tracking, auditing and verifying the production stages. The FSC is too complex because of the importance of food security applications, waste reduction and various players such as producers, transportation companies, distributors, retailers and end-users (Manupati et al. 2019). Blockchain technology can improve food origin traceability and create trust in the entire supply chain (Xiong et al. 2020).

10.5.4. *A mathematical model example in the blockchain concept*

Manupati et al. (2020) consider a three-echelon supply chain system in which raw materials and products flow from suppliers to distributors. The proposed model is solved by leveraging a blockchain approach to optimize the operational demands and the carbon emissions, the quantities of carbon emissions at each outlet act as the transaction medium. The amount of carbon emissions (carbon credits) is denoted by our system's carbon assets and is continuously transferred from the sender's account to the receiver's account in line with the transfer of the goods. Smart contracts are used for each party to monitor every transaction and store carbon emissions data in the form of a decentralized distributed ledger of its outlet. Each outlet has a limit: if any sudden change is observed or if the threshold limit is surpassed, the smart contract triggers an alert stating that the emission in the corresponding outlet is unstable. The updated data from the respective outlet is taken as an input for optimizing the system using our proposed algorithm described below. Subsequently, changes in operational demands must be made to stabilize the system. At this point, the smart contracts will start the analysis by examining the emissions in the respective outlet, and the iterations will continue. The model variables and parameters are presented in Table 10.5.

Sets	
S	Supplier
M	Manufacturing plant
W	Distribution center
I	Item to be supplied to manufacturing plant M
P	Product delivered to distribution center W
D	Demand

Parameters and variables	
L_{MI}	Lead time
r_{MI}	Reorder point
$Z_{1-\alpha}$	Service level of the supply chain
σ_{LT}	Variance in demand during the lead time
H_{MI}	Holding cost of item I at plant M
Q_{MI}	Order quantity at plant M
O_{MI}	Ordering cost of I
S_{capM}	Supply limit for items to manufacturing plant M
P_{capMP}	Production limit for product P in manufacturing plant M
F_{M}	Fixed cost of operating plant M
μ_{WM}	Mean demand of product P
σ_{WM}	Variance in demand of product P
μ_{DWP}	Mean of customer demand at W for product P
σ_{DWP}	Variance of customer demand at W for product P
D_{MI}	Demand of item I in plant M
V_{MI}	Variance of item I in plant M
H_{WP}	Holding cost at distribution center P
QR_{WPMt}	Regular time production quantity of product P from M to W
QO_{WPMt}	Over-time production quantity of product P from M to W
C_{MPt}	Cost per unit of regular time production of product P at M
C_{WPMt}	Cost per unit of over-time production of product P at M
T_{MSI}	Transportation cost of item I from supplier S to plant M
T_{WMP}	Transportation cost of product P from plant M to distribution center W
E_{FM}	Fixed emissions from plant M
E_{VM}	Variable emissions from plant M
TE_{FPM}	Fixed emissions due to the transportation of product P from plant M
TE_{VPM}	Variable emissions due to the transportation of product P from plant M
TE_{FIS}	Fixed emissions due to the transportation of item I from supplier S
TE_{VIS}	Variable emissions due to the transportation of item I from supplier S
EI_{PW}	Emissions due to inventory at distribution center W

EI_{PMt}	Emissions due to inventory at plant M for time t
T	Carbon tax
Ψ	Trading cost of carbon credits
X_M	Binary variable that is 0: plant M is open; 1: plant M is closed
Y_{WMP}	Binary variable that is 0: material is transported to warehouse W from plant M; 1: material is transported to warehouse W from plant M
Z_{MSI}	Binary variable that is 0: supplier S serves item I to plant M; 1: supplier does not serve plant M
S	Binary variable that is 0: smart contract is triggered; 1: smart contract is not triggered
Z	Total operating cost

Table 10.5. *Notations, parameters and decision variables*

Manupati et al. (2020) developed a mixed-integer nonlinear programming (MINLP) model for calculating the total optimal cost with emissions, considering inventory and safety stocks and leveraging blockchain technology. The objective is to minimize the total cost and emissions at each outlet. The objective function is built by defining assumptions, parameters and decision variables of the model. The objective function with these constraints is then solved with the following assumptions:

1) The lead time of item I at plant M is a constant parameter.

2) The entire SCN parameters have a standard normal distribution.

3) There are no constraints on order quantity, and capacity limitations are neglected.

4) Demands of participants in the supply chain are fulfilled.

5) Assembly times include setup times. Since the demands are fulfilled, there is no possibility of an item shortage while assembling the final product.

6) An output product, which may either be an intermediate or end product, is manufactured by facility M (M ∈ M1, M2, M3). Each section of this output product may be obtained from multiple sections of distinct input products.

7) Facility M may have several nodes towards the source, including other manufacturing sectors or external suppliers. It is assumed that each facility M has one node towards the source for each input product.

8) The emission rate is not fixed and is monitored for each outlet.

9) The transportation emission between any two outlets is considered as the emissions of the initial location (i.e. if the goods are transferred from supplier S to manufacturer M, the transportation emission encountered during the process is the emission of the supplier S).

10) For ease of interpretation, we assume that the smart contracts are triggered in the case study (i.e. S = 1).

The proposed algorithm is presented in equation (1) and consists of several parts. The first part of the objective function indicates the fixed total costs with fixed emissions. The second part refers to the ordering and total holding cost at each manufacturing site based on the economic order quantity (EOQ). The third part specifies the calculations based on the reorder point, including any buffer or safety stock. The fourth part accounts for the inventory holding cost along with the emission cost (where emission cost is the product of emission quantity and carbon taxation based on the policy). The fifth part refers to the total manufacturing and emission costs due to manufacturing. The sixth part signifies the transportation cost from S to M and its associated emission cost. The seventh part represents the transportation cost from M to W and its associated emission cost:

$$Z = S * \begin{bmatrix} \sum\limits_{M} F_M X_M + E_{FM} X_M \tau + \sum\limits_{M}\sum\limits_{I} \sqrt{2(H_{MI} + EI_{PMt}\tau)O_{MI}}\sqrt{D_{MI}} \\ + \sum\limits_{M}\sum\limits_{I} (H_{MI} + EI_{PMt}\tau) Z_{1-\alpha}\sqrt{L_{MI}}\sqrt{V_{MI}} + \sum\limits_{C}\sum\limits_{P}\sum\limits_{t} \mu_{WM}(H_{WP} + EI_{PW}\tau) \\ + \sum\limits_{W}\sum\limits_{M}\sum\limits_{P}\sum\limits_{T} (QR_{WMPt}C_{MPt} + QO_{WPMt}C_{WPMt}) + \sum\limits_{M=1} E_{VM}(QR_{WPMt} + QO_{WPMt})\tau \\ + \sum\limits_{M}\sum\limits_{S}\sum\limits_{I} T_{MSI}D_{MI} + Z_{MSI} + (TE_{FIS}Z_{MSI} + TE_{VIS}D_{MI})\tau \\ + \sum\limits_{W}\sum\limits_{M}\sum\limits_{P} T_{WMP}\mu_{WM}Y_{WMP} + (TE_{FMP}Y_{WMP} + TE_{VMP}\mu_{WM})\tau \end{bmatrix} \quad [10.81]$$

Subject to:

$$\Pr(D(L_{MI}) \leq r_{MI}) = 1 - \alpha \quad [10.82]$$

$$r_{MI} = E(D_{MI}) * E(L_{MI}) + Z_{1-\alpha}\sqrt{E(L_{MI}V_{MI}) + E(D_{MI})^2 \sigma_{LT}^2} \quad [10.83]$$

$$r_{MI} = D_{MI}E(L_{MI}) + Z_{(1-\alpha)}\sqrt{V_{MI}L_{MI}} \quad [10.84]$$

$$H_{MI}(Q_{MI} / 2) + H_{MI}Z_{1-\alpha}\sqrt{L_{MI}V_{MI}} \quad [10.85]$$

$$\sum_{M}\sum_{I} O_{MI}(D_{MI} / Q_{MI}) + H_{MI}(Q_{MI} / 2) + \sum_{M}\sum_{I} H_{MI} Z_{1-\alpha}\sqrt{L_{MI}V_{MI}} \quad [10.86]$$

$$Q_{MI} = \sqrt{\frac{2O_{MI}D_{MI}}{H_{MI}}} \quad [10.87]$$

$$\sum_{S} Z_{MSI} = X_M \;\; \forall I = 1,...,I \,\&\, M = 1,...,M \quad [10.88]$$

$$\sum_{M} Y_{WMP} = 1 \;\; \forall W = 1,...,W \,\&\, P = 1,...,P \quad [10.89]$$

$$\sum_{S}\sum_{I} D_{MI} S_I Z_{MSI} \leq S_{capM} X_M \;\; \forall M = 1,...,M \quad [10.90]$$

$$\sum_{W}\sum_{P} \mu_{WP} T_P Y_{WMP} \leq P_{capMP} \;\; \forall M = 1,...,M \quad [10.91]$$

$$\sum_{W}\sum_{I} \mu_{WP} Y_{WMP} b_{PI} \leq D_{MI} \;\; \forall M = 1,...,M \,\&\, P = 1,...,P \quad [10.92]$$

$$\sum_{W}\sum_{I} \sigma_{WP} Y_{VMP} b_{PI}^2 = V_{MI} \;\; \forall M = 1,...,M \,\&\, P = 1,...,P \quad [10.93]$$

$$L_{nWP(t-1)} + QR_{WMPt} = L_{nPt} \;\; \forall W, M, P \quad [10.94]$$

$$\sum_{W}\sum_{P} QR_{WMPt} T_P \leq T_{RMt} \;\; \forall M = 1,...,M \quad [10.95]$$

$$\sum_{W}\sum_{P} QO_{WMPt} T_p \leq T_{OMt} \;\; \forall M \quad [10.96]$$

$$\sum_{W} L_{nWPt} U_P \leq S_{capW} \;\; \forall W, t \quad [10.97]$$

$$\sum_{t} QR_{WMPt} + QO_{WMPt} \leq Y_{WMPt} \;\; \forall W, M, P \quad [10.98]$$

$$QR_{WMPt}, QO_{WMPt} \geq 0 \quad [10.99]$$

$$X_M, Y_{WMP}, Z_{MSI} \in \{0,1\} \quad [10.100]$$

The function is to reduce the total optimal cost and carbon emissions of the entire system. The model accounts for inventory, transportation, ordering and manufacturing costs. Both regular and overtime manufacturing costs are considered a safety stock and the respective emissions. The inventory is stockpiled in the distribution center to satisfy customer demand with an expected probability of (1-α) during the lead time L_{MI}. This probability function is determined by inventory service levels as presented in equation [10.82]. The reorder point is calculated considering the safety stock and assuming a normal distribution. This is reflected in equation [10.83]. Here, σ_{LT} is the variance for the demand in the lead time. While considering a constant lead time, we can neglect the variance and the reorder point in equation [10.84]. Here, Z1−α is indicated as the standard normal distribution value, which is uniform across the network. Based on equation [10.85], the inventory holding cost is calculated by equation [10.85], which includes information on the average holding cost, the EOQ and the cost incurred by holding safety stock. Therefore, equation [10.86] represents the total holding and ordering costs. According to the third assumption noted above, capacity limitations are not considered. Setting equation [10.86] to zero yields an equation in terms of Q, which results in equation [10.87]. Combining equations [10.86] and [10.87], the aggregate production allocation model is formulated to decrease the total optimal cost of the supply chain system. Here, equation [10.88] ensures that the facility of supplier S is open such that it provides items I to satisfy the demands of manufacturing unit M. Equation [10.89] ensures that all distribution center (warehouse) demands are fulfilled for a single operating plant, and Y_{WMP} states that products P are transported from manufacturing plants M to warehouses W. Constraints [10.90] and [10.91] denote the capacity shortage and production limitations, respectively, encountered by manufacturing plant M. Equations [10.92] and [10.93] are applied to generate the mean and variance, respectively, for product P, which is to be manufactured at plant M. Constraint [10.94] is used to balance the demands for each distribution center by considering inventories of the current and the previous period. It also balances the quantity to be produced for each product P. Equations [10.95] and [10.96] state the production quantity constraints for regular and overtime manufacturing hours, respectively. Equation [10.97] reflects the storage capacity of the warehouse. Constraint [10.98] ensures that product P is produced at opened plant M. In equation [10.99], X, Y and Z are binary variables, taking on values of either 0 or 1. Constraint [10.100] verifies that regular time and overtime production are always greater than zero.

10.6. References

Alam, S.S. and Sayuti, N.M. (2011). Applying the theory of planned behavior (TPB) in halal food purchasing. *International Journal of Commerce and Management*, 21(1), 8–20.

Alcamo, J., Flörke, M., Märker, M. (2007). Future long-term changes in global water resources driven by socio-economic and climatic changes. *Hydrological Sciences Journal*, 52(2), 247–275.

Aras, N. and Bilge, Ü. (2018). Robust supply chain network design with multi-products for a company in the food sector. *Applied Mathematical Modelling*, 60, 526–539.

Assen, M.F. (2010). *Praktijkboek supply chain management: Aanpak, concepten en modellen voor operational excellence in de keten*. Kluwer, Alphen aan den Rijn.

Athauda, T. and Karmakar, N.C. (2019). Review of RFID-based sensing in monitoring physical stimuli in smart packaging for food-freshness applications. *Wireless Power Transfer*, 6(2), 161–174.

Banasik, A., Kanellopoulos, A., Claassen, G.D.H., Bloemhof-Ruwaard, J.M., van der Vorst, J.G. (2017). Closing loops in agricultural supply chains using multi-objective optimization: A case study of an industrial mushroom supply chain. *International Journal of Production Economics*, 183, 409–420.

Barrett, C.B. and Mutambatsere, E. (2008). Agricultural markets in developing countries. In *The New Palgrave Dictionary of Economics*. Palgrave Macmillan, London.

Battini, D., Bogataj, M., Choudhary, A. (2017). Closed loop supply chain (CLSC): Economics, modelling, management and control. *International Journal of Production Economics*, 183, 319–321.

Beamon, B.M. (1998). Supply chain design and analysis: Models and methods. *International Journal of Production Economics*, 55(3), 281–294.

Ben-Daya, M., Hassini, E., Bahroun, Z. (2019). Internet of things and supply chain management: A literature review. *International Journal of Production Research*, 57(15–16), 4719–4742.

Bottani, E., Murino, T., Schiavo, M., Akkerman, R. (2019). Resilient food supply chain design: Modelling framework and metaheuristic solution approach. *Computers & Industrial Engineering*, 135, 177–198.

Brewster, C., Roussaki, I., Kalatzis, N., Doolin, K., Ellis, K. (2017). IoT in agriculture: Designing a Europe-wide large-scale pilot. *IEEE Communications Magazine*, 55(9), 26–33.

Carter, C.R. and Ellram, L.M. (1998). Reverse logistics: A review of the literature and framework for future investigation. *Journal of Business Logistics*, 19(1), 85.

Chaabane, A., Ramudhin, A., Paquet, M. (2012). Design of sustainable supply chains under the emission trading scheme. *International Journal of Production Economics*, 135(1), 37–49.

Chao, C., Zhihui, T., Baozhen, Y. (2019). Optimization of two-stage location–routing–inventory problem with time-windows in food distribution network. *Annals of Operations Research*, 273, 111–134.

Cheraghalipour, A., Paydar, M.M., Hajiaghaei-Keshteli, M. (2018). A bi-objective optimization for citrus closed-loop supply chain using Pareto-based algorithms. *Applied Soft Computing*, 69, 33–59.

Chopra, S. (2003). Designing the distribution network in a supply chain. *Transportation Research Part E: Logistics and Transportation Review*, 39(2), 123–140.

Crosby, M., Pattanayak, P., Verma, S., Kalyanaraman, V. (2016). Blockchain technology: Beyond bitcoin. *Applied Innovation*, 2(6–10), 71.

Cui, L., Guo, S., Zhang, H. (2020). Coordinating a green agri-food supply chain with revenue-sharing contracts considering retailers' green marketing efforts. *Sustainability*, 12(4), 1289.

Dadi, V., Nikhil, S.R., Mor, R.S., Agarwal, T., Arora, S. (2021). Agri-food 4.0 and innovations: Revamping the supply chain operations. *Production Engineering Archives*, 27(2), 75–89.

Dahane, A., Benameur, R., Kechar, B. (2022). An IoT low-cost smart farming for enhancing irrigation efficiency of smallholder's farmers. *Wireless Personal Communications*, 127(4), 3173–3210.

Dania, W.A.P., Xing, K., Amer, Y. (2016). Collaboration and sustainable agri-food suply chain: A literature review. *MATEC Web of Conferences*, 58, 02004.

Darbari, J.D., Kannan, D., Agarwal, V., Jha, P.C. (2019). Fuzzy criteria programming approach for optimising the TBL performance of closed loop supply chain network design problem. *Annals of Operations Research*, 273, 693–738.

Dauvergne, P. and Lister, J. (2012). Big brand sustainability: Governance prospects and environmental limits. *Global Environmental Change*, 22(1), 36–45.

Dekker, R., Fleischmann, M., Inderfurth, K., van Wassenhove, L.N. (eds) (2013). *Reverse Logistics: Quantitative Models for Closed-Loop Supply Chains*. Springer Science & Business Media, Berlin/Heidelberg.

Dyckhoff, H., Lackes, R., Reese, J. (eds) (2013). *Supply Chain Management and Reverse Logistics*. Springer Science & Business Media, Berlin/Heidelberg.

Eksoz, C., Mansouri, S.A., Bourlakis, M. (2014). Collaborative forecasting in the food supply chain: A conceptual framework. *International Journal of Production Economics*, 158, 120–135.

Ellis, S., Morris, H.D., Santagate, J. (2015). IoT-enabled analytic applications revolutionize supply chain planning and execution. White Paper, 13, International Data Corporation (IDC), Needham.

Fadhel, A.W. (2021). Closed-loop sustainable food supply chain management: Design of network models for efficient operations. PhD Thesis, Northeastern University, Boston.

Fang, L., Bai, Z., Wei, S., Yanfen, H., Zongming, W., Kaishan, S., Zhiming, L. (2008). Sandy desertification change and its driving forces in western Jilin Province, North China. *Environmental Monitoring and Assessment*, 136(1), 379–390.

Fazlollahtabar, H. (2018). *Supply Chain Management Models: Forward, Reverse, Uncertain, and Intelligent Foundations with Case Studies*. CRC Press, Boca Raton.

Flapper, S.D., van Nunen, J., Van Wassenhove, L.N. (eds) (2005). *Managing Closed-loop Supply Chains*. Springer Science & Business Media, Berlin/Heidelberg.

Fleischmann, M., Bloemhof-Ruwaard, J.M., Dekker, R., Van der Laan, E., Van Nunen, J.A., Van Wassenhove, L.N. (1997). Quantitative models for reverse logistics: A review. *European Journal of Operational Research*, 103(1), 1–17.

Gharehgozli, A., Iakovou, E., Chang, Y., Swaney, R. (2017). Trends in global E-food supply chain and implications for transport: Literature review and research directions. *Research in Transportation Business & Management*, 25, 2–14.

Gkountani, V.A., Tsoulfas, G.T., Rachaniotis, N.P. (2021). Circular economy and resilience: Convergences and deviations in the case of agri-food supply chains. In *IOP Conference Series: Earth and Environmental Science*, 23–24 October, 899(1), 012001. IOP Publishing, Athens.

Godfray, H.C.J., Beddington, J.R., Crute, I.R., Haddad, L., Lawrence, D., Muir, J.F., Pretty, J., Robinson, S., Thomas, S.M., Toulmin, C. (2010). Food security: The challenge of feeding 9 billion people. *Science*, 327(5967), 812–818.

Goedde, L., Horii, M., Sanghvi, S. (2015). *Pursuing the Global Opportunity in Food and Agribusiness*. McKinsey Global Institute, Washington, DC.

Gogo, E.O., Opiyo, A.M., Ulrichs, C., Huyskens-Keil, S. (2017). Nutritional and economic postharvest loss analysis of African indigenous leafy vegetables along the supply chain in Kenya. *Postharvest Biology and Technology*, 130, 39–47.

Govindan, K., Jafarian, A., Khodaverdi, R., Devika, K. (2014). Two-echelon multiple-vehicle location–routing problem with time windows for optimization of sustainable supply chain network of perishable food. *International Journal of Production Economics*, 152, 9–28.

Govindan, K., Soleimani, H., Kannan, D. (2015). Reverse logistics and closed-loop supply chain: A comprehensive review to explore the future. *European Journal of Operational Research*, 240(3), 603–626.

Govindan, K., Cheng, T.E., Mishra, N., Shukla, N. (2018). Big data analytics and application for logistics and supply chain management. *Transportation Research Part E: Logistics and Transportation Review*, 114, 343–349.

Govindan, K., Mina, H., Esmaeili, A., Gholami-Zanjani, S.M. (2020). An integrated hybrid approach for circular supplier selection and closed loop supply chain network design under uncertainty. *Journal of Cleaner Production*, 242, 118317.

Guide, V.D.R., Harrison, T.P., Van Wassenhove, L.N. (2003). The challenge of closed-loop supply chains. *Interfaces*, 33(6), 3–6.

Heck, S., Campos, H., Barker, I., Okello, J.J., Baral, A., Boy, E., Birol, E. (2020). Resilient agri-food systems for nutrition amidst COVID-19: Evidence and lessons from food-based approaches to overcome micronutrient deficiency and rebuild livelihoods after crises. *Food Security*, 12(4), 823–830.

Hernández, J.E., Kacprzyk, J., Lyons, A., Ortiz, A., Panetto, H. (2018). Review on operational research advances in agri-food supply chains and societal challenges. In *29th European Conference on Operational Research, EURO'2018*. RUC-APS, Valencia.

Hobbs, J.E. (2020). Food supply chains during the COVID-19 pandemic. *Canadian Journal of Agricultural Economics*, 68(2), 171–176.

Hoellthaler, G., Meister, F., Braunreuther, S., Reinhart, G. (2020). Function framework for describing digital technologies in the context of lean production. *Procedia CIRP*, 88(7), 167–172.

Jayaraman, V. and Luo, Y. (2007). Creating competitive advantages through new value creation: A reverse logistics perspective. *Academy of Management Perspectives*, 21(2), 56–73.

Jia, X., Feng, Q., Fan, T., Lei, Q. (2012). RFID technology and its applications in Internet of Things (IoT). In *2nd International Conference on Consumer Electronics, Communications and Networks, (CECNet)*. IEEE, Yichang, 1282–1285.

Jiang, Y., Zhao, L., Sun, S. (2009). A resilient strategy for meat-food supply chain network design. In *IEEE International Conference on Industrial Engineering and Engineering Management*, Hong Kong, 1479–1483.

Kadambala, D.K., Subramanian, N., Tiwari, M.K., Abdulrahman, M., Liu, C. (2017). Closed loop supply chain networks: Designs for energy and time value efficiency. *International Journal of Production Economics*, 183, 382–393.

Kalantari, F. and Hosseininezhad, S.J. (2022). A multi-objective cross entropy-based algorithm for sustainable global food supply chain with risk considerations: A case study. *Computers & Industrial Engineering*, 164, 107766.

Kamble, S.S., Gunasekaran, A., Gawankar, S.A. (2020). Achieving sustainable performance in a data-driven agriculture supply chain: A review for research and applications. *International Journal of Production Economics*, 219, 179–194.

Kamilaris, A., Fonts, A., Prenafeta-Boldú, F.X. (2019). The rise of blockchain technology in agriculture and food supply chains. *Trends in Food Science & Technology*, 91, 640–652.

Keramydas, C., Papapanagiotou, K., Vlachos, D., Iakovou, E. (2015). Risk management for agri-food supply chains. In *Supply Chain Management for Sustainable Food Networks*, Iakovou, E., Bochtis, D., Vlachos, D., Aidonis, D. (eds). John Wiley & Sons, Hoboken, 255–292.

Kittipanya-Ngam, P. and Tan, K.H. (2020). A framework for food supply chain digitalization: Lessons from Thailand. *Production Planning & Control*, 31(2–3), 158–172.

Klitou, D., Conrads, J., Rasmussen, M., Probst, L., Pedersen, B. (2017). Digital Transformation Monitor. Germany: Industrie 4.0. Document, European Commission.

Kotler, P. and Pfoertsch, W. (2010). *Ingredient Branding: Making the Invisible Visible.* Springer Science & Business Media, Berlin/Heidelberg.

Kotler, P., Kartajaya, H., Setiawan, I. (2019). *Marketing 3.0: From Products to Customers to the Human Spirit.* Springer, Singapore.

Kumar, P. and Kumar Singh, R. (2022). Strategic framework for developing resilience in agri-food supply chains during COVID-19 pandemic. *International Journal of Logistics Research and Applications*, 25(11), 1401–1424.

Kumar, V. and Srivastava, A. (2018). The role of RFID in agro-food sector, agricultural research & technology. *Open Access Journal*, 14(4), 1–7.

Kumar, A., Mangla, S.K., Kumar, P. (2022). An integrated literature review on sustainable food supply chains: Exploring research themes and future directions. *Science of the Total Environment*, 153411.

Kwag, S.I. and Ko, Y.D. (2019). Optimal design for the halal food logistics network. *Transportation Research Part E: Logistics and Transportation Review*, 128, 212–228.

Lambert, S.R., Elamin, N.E.A., De Cordoba, S.F. (2020). Maximizing sustainable agri-food supply chain opportunities to redress COVID-19 in developing countries. Document, United Nations Conference on Trade and Development.

Lezoche, M., Hernandez, J.E., Díaz, M.D.M.E.A., Panetto, H., Kacprzyk, J. (2020). Agri-food 4.0: A survey of the supply chains and technologies for the future agriculture. *Computers in Industry*, 117, 103187.

Li, D., Wang, X., Chan, H.K., Manzini, R. (2014). Sustainable food supply chain management. *International Journal of Production Economics*, 152, 1–8.

Li, Y., Dai, J., Cui, L. (2020). The impact of digital technologies on economic and environmental performance in the context of industry 4.0: A moderated mediation model. *International Journal of Production Economics*, 229, 107777.

Liu, X.L. and Ji, J. H. (2007). Research on resilient supply network based on multi-sourcing. *Industrial Engineering and Management*, 12(3), 8–12.

Mabert, V.A. and Venkataramanan, M.A. (1998). Special research focus on supply chain linkages: Challenges for design and management in the 21st century. *Decision Sciences*, 29(3), 537–552.

Mahbub, M. (2020). A smart farming concept based on smart embedded electronics, internet of things and wireless sensor network. *Internet of Things*, 9, 1–30.

Mangla, S.K., Govindan, K., Luthra, S. (2016). Critical success factors for reverse logistics in Indian industries: A structural model. *Journal of Cleaner Production*, 129, 608–621.

Manupati, V.K., Jedidah, S.J., Gupta, S., Bhandari, A., Ramkumar, M. (2019). Optimization of a multi-echelon sustainable production-distribution supply chain system with lead time consideration under carbon emission policies. *Computers & Industrial Engineering*, 135, 1312–1323.

Manupati, V.K., Schoenherr, T., Ramkumar, M., Wagner, S.M., Pabba, S.K., Inder Raj Singh, R. (2020). A blockchain-based approach for a multi-echelon sustainable supply chain. *International Journal of Production Research*, 58(7), 2222–2241.

Mardan, E., Govindan, K., Mina, H., Gholami-Zanjani, S.M. (2019). An accelerated benders decomposition algorithm for a bi-objective green closed loop supply chain network design problem. *Journal of Cleaner Production*, 235, 1499–1514.

Mavi, R.K., Goh, M., Zarbakhshnia, N. (2017). Sustainable third-party reverse logistic provider selection with fuzzy SWARA and fuzzy MOORA in plastic industry. *The International Journal of Advanced Manufacturing Technology*, 91(5), 2401–2418.

McLeod, S. (2007). Maslow's hierarchy of needs. *Simply Psychology*, 1(1–18).

Mohammed, A. and Wang, Q. (2017). Multi-criteria optimization for a cost-effective design of an RFID-based meat supply chain. *British Food Journal*, 119(3), 676–689.

Myers, R.J., Sexton, R.J., Tomek, W.G. (2010). A century of research on agricultural markets. *American Journal of Agricultural Economics*, 92(2), 376–403.

Narassima, M.S., Anbuudayasankar, S.P., Mathiyazhagan, K., Ganesh, K., Lee, T.R. (2024). Supply chain resilience: Conceptual model building and validation. *International Journal of Logistics Research and Applications*, 27(5), 781–813.

O'Byrne, R. (2020). 6 supply chain trends from 2017 and their ongoing progress. Logistics Bureau [Online]. Available at: https://www.logisticsbureau.com/6-key-supply-chain-and-logistics-trends-to-watch-in-2017.

OECD (2019). The Digital Transformation of the Agriculture and Food System. Agriculture policy brief.

OECD (2020). OECD scheme for the application of international standards for fruit and vegetables. Brochure, Committee for Agriculture, TAD/CA/FVS/WD(2020)1/REV7 [Online]. Available at: https://www.oecd.org/agriculture/fruit-vegetables/oecd-covid-19-impact-on-fruit-and-vegetables-trade.pdf.

Othman, R., Ahmad, Z.A., Zailani, S. (2009). The effect of institutional pressures in the Malaysian halal food industry. *International Business Management*, 3(4), 80–84.

Ozsen, L., Coullard, C.R., Daskin, M.S. (2008). Capacitated warehouse location model with risk pooling. *Naval Research Logistics (NRL)*, 55(4), 295–312.

Paksoy, T., Bektaş, T., Özceylan, E. (2011). Operational and environmental performance measures in a multi-product closed-loop supply chain. *Transportation Research Part E: Logistics and Transportation Review*, 47(4), 532–546.

Prakash, C. and Barua, M.K. (2016). A multi-criteria decision-making approach for prioritizing reverse logistics adoption barriers under fuzzy environment: Case of Indian electronics industry. *Global Business Review*, 17(5), 1107–1124.

Queiroz, M.M. and Wamba, S.F. (2019). Blockchain adoption challenges in supply chain: An empirical investigation of the main drivers in India and the USA. *International Journal of Information Management*, 46, 70–82.

Ray, P. (2021). Agricultural supply chain risk management under price and demand uncertainty. *International Journal of System Dynamics Applications, (IJSDA)*, 10(2), 17–32.

Reyhani Yamchi, H., Jabarzadeh, Y., Ghaffarinasab, N., Kumar, V., Garza-Reyes, J.A. (2020). A multi-objective linear optimization model for designing sustainable closed-loop agricultural supply chain. In *Proceedings of the International Conference on Industrial Engineering and Operations Management*, 10–12 March. IEOM Society International, Dubai.

Robinson, M. (2017). Tiny self-driving robots have started delivering food on-demand in Silicon Valley – take a look. *Business Insider* [Online]. Available at: https://www.businessinsider.in/thelife/tiny-self-driving-robots-have-started-delivering-food-on-demand-in-silicon-valley-take-a-look/slidelist/57817087.cms.

Roßmann, B., Canzaniello, A., von der Gracht, H., Hartmann, E. (2018). The future and social impact of Big Data Analytics in Supply Chain Management: Results from a Delphi study. *Technological Forecasting and Social Change*, 130, 135–149.

Rogers, D. and Tibben-Lembke, R. (2001). An examination of reverse logistic practices. *Journal of Business Logistics*, 22(2), 129–148.

Rohmer, S.U.K., Gerdessen, J.C., Claassen, G.D.H. (2019). Sustainable supply chain design in the food system with dietary considerations: A multi-objective analysis. *European Journal of Operational Research*, 273(3), 1149–1164.

Rueda, X., Garrett, R.D., Lambin, E.F. (2017). Corporate investments in supply chain sustainability: Selecting instruments in the Agri-food industry. *Journal of Cleaner Production*, 142, 2480–2492.

de Sá, M.M., de Souza Miguel, P.L., de Brito, R.P., Pereira, S.C.F. (2019). Supply chain resilience: The whole is not the sum of the parts. *International Journal of Operations & Production Management*, 40(1), 92–115.

Saberi, S., Kouhizadeh, M., Sarkis, J., Shen, L. (2019). Blockchain technology and its relationships to sustainable supply chain management. *International Journal of Production Research*, 57(7), 2117–2135.

Santos, C., Mehrsai, A., Barros, A.C., Araújo, M., Ares, E. (2017). Towards Industry 4.0: An overview of European strategic roadmaps. *Procedia Manufacturing*, 13, 972–979.

Shah Alam, S. and Mohamed Sayuti, N. (2011). Applying the Theory of Planned Behavior (TPB) in halal food purchasing. *International Journal of Commerce and Management*, 21(1), 8–20.

Shokri, A., Oglethorpe, D., Nabhani, F. (2014). Evaluating sustainability in the UK fast food supply chain: Review of dimensions, awareness and practice. *Journal of Manufacturing Technology Management*, 25(8), 1224–1244.

Shukla, M. and Jharkharia, S. (2013). Agri-fresh produce supply chain management: A state-of-the-art literature review. *International Journal of Operations & Production Management*, 33(2), 114–158.

Soosay, C. and Kannusamy, R. (2018). Scope for industry 4.0 in agri-food supply chain. The road to a digitalized supply chain management: Smart and digital solutions for supply chain management. In *Proceedings of the Hamburg International Conference of Logistics (HICL)*, 25, 37–56. Hamburg University of Technology (TUHH) and Institute of Business Logistics and General Management, Berlin.

Souza, V.D., Bloemhof-Ruwaard, J., Borsato, M. (2019). Exploring ecosystem network analysis to balance resilience and performance in sustainable supply chain design. *International Journal of Advanced Operations Management*, 11(1–2), 26–45.

Strecker, K., Bitzer, V., Kruijssen, F. (2022). Critical stages for post-harvest losses and nutrition outcomes in the value chains of bush beans and nightshade in Uganda. *Food Security*, 14(2), 411–426.

Thyberg, K.L. and Tonjes, D.J. (2016). Drivers of food waste and their implications for sustainable policy development. *Resources, Conservation and Recycling*, 106, 110–123.

Tijskens, L.M.M., Hertog, M.L.A.T.M., Nicolaï, B.M. (eds) (2001). *Food Process Modelling*. Woodhead Publishing, Cambridge.

Trienekens, J.H., Wognum, P.M., Beulens, A.J., van der Vorst, J.G. (2012). Transparency in complex dynamic food supply chains. *Advanced Engineering Informatics*, 26(1), 55–65.

Tsai, C.W., Lai, C.F., Chao, H.C., Vasilakos, A.V. (2015). Big data analytics: A survey. *Journal of Big Data*, 2(1), 1–32.

UNCTAD (2020a). Global trade update [Online]. Available at: https://unctad.org/en/PublicationsLibrary/ditcmisc2020d2_en.pdf.

Verdouw, C.N., Wolfert, J., Beulens, A.J.M., Rialland, A. (2016). Virtualization of food supply chains with the internet of things. *Journal of Food Engineering*, 176, 128–136.

Xiao-hui, Y. and Jun, M. (2012). Development of sustainable food supply chain. *International Journal of Engineering and Technology*, 4(2), 166.

Xiong, H., Dalhaus, T., Wang, P., Huang, J. (2020). Blockchain technology for agriculture: Applications and rationale. *Frontiers in Blockchain*, 3, 7.

Yadav, V.S., Singh, A.R., Gunasekaran, A., Raut, R.D., Narkhede, B.E. (2022). A systematic literature review of the agro-food supply chain: Challenges, network design, and performance measurement perspectives. *Sustainable Production and Consumption*, 29, 685–704.

Yousef, D.K. (2010). UAE: Halal food numbers look tasty. Size of global Muslim population creates significant customerbase [Online]. Available at: https://gulfnews.com/business/halal-food-numbers-look-tasty-1.679007.

Zakaria, N. and Abdul-Talib, A.N. (2010). Applying Islamic market-oriented cultural model to sensitize strategies towards global customers, competitors, and environment. *Journal of Islamic Marketing*, 1(1), 51–62.

Zandkarimkhani, S., Mina, H., Biuki, M., Govindan, K. (2020). A chance constrained fuzzy goal programming approach for perishable pharmaceutical supply chain network design. *Annals of Operations Research*, 295(1), 425–452.

Zhao, R., Liu, Y., Zhang, N., Huang, T. (2017). An optimization model for green supply chain management by using a big data analytic approach. *Journal of Cleaner Production*, 142, 1085–1097.

Zhu, Z., Chu, F., Dolgui, A., Chu, C., Zhou, W., Piramuthu, S. (2018). Recent advances and opportunities in sustainable food supply chain: A model-oriented review. *International Journal of Production Research*, 56(17), 5700–5722.

List of Authors

Muhammad Bilal AHSAN
Department of Agribusiness and Entrepreneurship Development
Muhammad Nawaz Sharif University of Agriculture
Multan
Pakistan

Mohamed Haykal AMMAR
OLID Research Laboratory
Higher Institute of Industrial Management
University of Sfax
Tunisia

Erick R. BANDALA
Desert Research Institute
Las Vegas
USA

Deepak BANGWAL
School of Business
University of Petroleum and Energy Studies (UPES)
Dehradun
India

Amina BENDJELLOUL
Manufacturing Engineering Laboratory of Tlemcen
Abou Bekr Belkaid University of Tlemcen
Algeria

Negin BERJIS
Department of Industrial Engineering
Najafabad Branch
Islamic Azad University
Najafabad
Iran

Piotr F. BOROWSKI
Vistula University
Warsaw
Poland

Omid BOYER
Department of Industrial Engineering
Najafabad Branch
Islamic Azad University
Najafabad
Iran

Pradeep CHAUHAN
School of Business
University of Petroleum and Energy Studies (UPES)
Dehradun
India

Mohamed Amine CHERIER
Manufacturing Engineering Laboratory of Tlemcen
Abou Bekr Belkaid University of Tlemcen
Algeria

Diala DHOUIB
ISGIS, OLID
University of Sfax
Tunisia

Chandra Prakash GARG
Department of Operations Management and Quantitative Techniques
Indian Institute of Management
Rohtak
Haryana
India

Zainab Alaa HAMEED
MODILS Laboratory
Faculty of Economics and Management of Sfax
University of Sfax
Tunisia

Githa S. HEGGDE
Mudra Institute of Communications Ahmedabad (MICA)
India

Rémy HOUSSIN
ICUBE
University of Strasbourg
INSA
France

Zohaib JABBAR
Institute of Business Management Sciences
University of Agriculture
Faisalqbad
Pakistan

Raoudha KAMMOUN
MODILS Laboratory
Faculty of Economics and Management of Sfax
University of Sfax
Tunisia

Farshad KAVEH
Department of Industrial Engineering
Najafabad Branch
Islamic Azad University
Najafabad
Iran

Mohamed Amir KESKES
LIMOS UMR CNRS 6158
Université Clermont Auvergne
Aubière
France
and
ISGIS, OLID
University of Sfax
Tunisia

Mayssa KOUBAA
OLID Research Laboratory
Higher Institute of Industrial Management
University of Sfax
Tunisia

Rupesh KUMAR
Jindal Global Business School (JGBS)
O.P. Jindal Global University
Sonipat
Haryana
India

Kailash Chandra LACHHWANI
Department of Applied Science
NITTTR
Chandigarh
India

Taicir Moalla LOUKIL
MODILS Laboratory
Faculty of Economics and Management of Sfax
University of Sfax
and
Tunisian Operational Research Society TORS
and
African Federation of Operational Research AFROS
Tunisia

Aleksandra LUBAŃSKA
Warsaw University of Life Sciences
Poland

Bouthaina MAAZOUN
MODILS Laboratory
Faculty of Economics and Management of Sfax
University of Sfax
Tunisia

Hafiz Zahid MAHMOOD
Department of Agribusiness and Entrepreneurship Development
Muhammad Nawaz Sharif University of Agriculture
Multan
Pakistan

Sirine MNEJJA
Ces Lab
National Engineering School of Sfax
University of Sfax
Tunisia

Deepanshu NAYAK
Department of Applied Mathematics and Scientific Computing
Indian Institute of Technology
Roorkee
India

Millie PANT
Department of Applied Mathematics and Scientific Computing
Indian Institute of Technology
Roorkee
India

Abd-Ur-REHMAN
Department of Agribusiness and Entrepreneurship Development
Muhammad Nawaz Sharif University of Agriculture
Multan
Pakistan

Jean RENAUD
ICUBE
University of Strasbourg
INSA
France
and
Département de Génie des Systèmes
ETS
Montreal
Canada

Nadia SAEIDI
Department of Industrial Engineering
Najafabad Branch
Islamic Azad University
Najafabad
Iran

Shila SHARIFZADEH
CEG-IST
Instituto Superior Técnico
University of Lisbon
Portugal

Archana SHARMA
Department of Applied Sciences
KIET Group of Institutions
Ghaziabad
India

Eric STEIN
Management Science and
Information Systems
Penn State University
Pennsylvania
USA

Munnaluru Sai Venkata YESWANTH
Department of Transportation
School of Business
University of Petroleum & Energy
Studies
Dehradun
India

Alaeddine ZOUARI
ISGIS, OLID
University of Sfax
Tunisia

Index

S, T, V, W

Other titles from

ISTE

in

Agriculture, Food Science and Nutrition

2024

CHERIET Foued, MAUREL Carole, AMADIEU Paul, HANNIN Hervé
Wine Management and Marketing 2: Responses of the Industry to Crises and New Expectations

2023

FLEURENCE Joël
Algae in Agrobiology: Realities and Perspectives

2020

CHERIET Foued, MAUREL Carole, AMADIEU Paul, HANNIN Hervé
Wine Management and Marketing: Opportunities for Companies and Challenges for the Industry

2019

YANNOU-LE BRIS Gwenola, SERHAN Hiam, DUCHAÎNE Sibylle, FERRANDI Jean-Marc, TRYSTRAM Gilles
Ecodesign and Ecoinnovation in the Food Industries

2016

JEANTET Romain, CROGUENNEC Thomas, SCHUCK Pierre, BRULÉ Gérard
Handbook of Food Science and Technology 1: Food Alteration and Food Quality
Handbook of Food Science and Technology 2: Food Process Engineering and Packaging
Handbook of Food Science and Technology 3: Food Biochemistry and Technology

Printed and bound by CPI Group (UK) Ltd, Croydon, CR0 4YY

28/07/2026

14931340-0004